T0321170

Seduced by Mathematics

The enduring fascination of mathematics

Problem Solving in Mathematics and Beyond

Print ISSN: 2591-7234
Online ISSN: 2591-7242

Series Editor: Dr. Alfred S. Posamentier
Distinguished Lecturer
New York City College of Technology - City University of New York

---

There are countless applications that would be considered problem solving in mathematics and beyond. One could even argue that most of mathematics in one way or another involves solving problems. However, this series is intended to be of interest to the general audience with the sole purpose of demonstrating the power and beauty of mathematics through clever problem-solving experiences.

Each of the books will be aimed at the general audience, which implies that the writing level will be such that it will not engulfed in technical language — rather the language will be simple everyday language so that the focus can remain on the content and not be distracted by unnecessarily sophiscated language. Again, the primary purpose of this series is to approach the topic of mathematics problem-solving in a most appealing and attractive way in order to win more of the general public to appreciate his most important subject rather than to fear it. At the same time we expect that professionals in the scientific community will also find these books attractive, as they will provide many entertaining surprises for the unsuspecting reader.

Published

Vol. 28 *Seduced by Mathematics: The Enduring Fascination of Mathematics*
by James D Stein

Vol. 27 *Mathematics: Its Historical Aspects, Wonders and Beyond*
by Alfred S Posamentier and Arthur D Kramer

Vol. 26 *Creative Secondary School Mathematics: 125 Enrichment Units for Grades 7 to 12*
by Alfred S Posamentier

Vol. 25 *Geometry in Our Three-Dimensional World*
by Alfred S Posamentier, Bernd Thaller, Christian Dorner, Robert Geretschläger, Guenter Maresch, Christian Spreitzer and David Stuhlpfarrer

For the complete list of volumes in this series, please visit www.worldscientific.com/series/psmb

Problem Solving in Mathematics and Beyond — Volume 28

Seduced by Mathematics

The enduring fascination of mathematics

James D Stein
California State University, USA

NEW JERSEY • LONDON • SINGAPORE • BEIJING • SHANGHAI • HONG KONG • TAIPEI • CHENNAI • TOKYO

Published by

World Scientific Publishing Co. Pte. Ltd.
5 Toh Tuck Link, Singapore 596224
USA office: 27 Warren Street, Suite 401-402, Hackensack, NJ 07601
UK office: 57 Shelton Street, Covent Garden, London WC2H 9HE

Library of Congress Control Number: 2022017162

British Library Cataloguing-in-Publication Data
A catalogue record for this book is available from the British Library.

Problem Solving in Mathematics and Beyond — Vol. 28
SEDUCED BY MATHEMATICS
The Enduring Fascination of Mathematics

ISBN 978-981-125-546-5 (hardcover)
ISBN 978-981-125-635-6 (paperback)
ISBN 978-981-125-547-2 (ebook for institutions)
ISBN 978-981-125-548-9 (ebook for individuals)

For any available supplementary material, please visit
https://www.worldscientific.com/worldscibooks/10.1142/12812#t=suppl

Desk Editors: Balasubramanian Shanmugam/Rok Ting Tan

Typeset by Stallion Press
Email: enquiries@stallionpress.com

Printed in Singapore

Preface — Just What Does She See in Him, Anyway?

Many of us know a bright, attractive woman with a good job and a lot going for her ... and a boyfriend who just doesn't seem to be in the same league. It's not that he's an unshaven slob but he doesn't seem to be the guy she deserves. And everyone has asked, "Just what does she see in him, anyway?"

"Just what do you see in math, anyway?" is a familiar question for everyone who likes, loves or teaches math — at least when they encounter someone who doesn't share their feelings. Of course, it's not just math. If you are a jazz aficionado or an American who likes soccer, you've probably heard something similar.

So what do you do? If you're a jazz aficionado, you find the most universally appealing jazz you can find and play it for your friend. If you like soccer, you find videos of incredible goals. But what do you do if you are me, and feel the way I do about math?

That's easy. You write this book.

This book would not have been written had I not encountered several people along the way who made critical contributions. First up is my wife, Linda. She and her mother visited Taiwan in early January 2020, and when she returned I contracted pneumonia. Despite my saying, "I'm sure it's just a cold," she made sure I got proper medical attention, including 6 weeks of caring for me when

I literally couldn't do anything. Covid was putting in an appearance at the time, and I'm pretty sure that had she not done what she did, I wouldn't be here to write this book.

Next are John Bachar and Ken Warner, two good friends and colleagues from California State University, Long Beach. Mathematicians are grouped in families in much the same way as real people are. Ken's thesis advisor at UCLA was Angus Taylor, who was also the thesis advisor for Bill Bade, who was my thesis advisor. John's thesis advisor was Phil Curtis, who co-authored many brilliant and insightful papers with Bill. Bill talked Phil into hiring me at UCLA, and John and Ken were instrumental in seeing that I was hired at CSULB and that I received tenure. I'm grateful to have this opportunity to express my appreciation for their support and friendship over the years.

Oh, yes, without John and Ken, I would not be married to the aforementioned Linda — because I met her while she was a grad student at CSULB. So, I owe them more than just a career.

And finally, there's Al Posamentier, with whom I have had the pleasure of doing a number of delightful podcasts on his excellent books, and who has enabled me to have my last three books published. Al and I have yet to meet in person, but I'm looking forward to the opportunity.

My father used to say of someone, "He's a gentleman and a scholar — and there are damned few of us left." I know John and Ken fall into that category, and I'd bet large sums that Al does, too.

About the Author

James D. Stein is a retired Professor of Mathematics from California State University (Long Beach, California), where he taught from 1975 to 2013. He graduated from Yale University with a B.A. in Mathematics and a minor in Physics, and received a Ph.D. in Mathematics from the University of California at Berkeley. He has published over 40 research papers, approximately half in Banach spaces and algebras and half in fixed points, but has also published several papers in physics and mathematics education. He taught courses at both undergraduate and graduate levels but specialized in teaching courses in mathematics education and calculus, real analysis and functional analysis. He served as a Content Review Expert for the textbook adoption for the State of California in both 2000 and 2013. He is also the author of approximately a dozen books on mathematics and science for the general public, including *The Fate of Schrodinger's Cat*, published by World Scientific Publishing. In retirement, he does podcasts with authors of books on mathematics and science for New Books Network.

Contents

Introduction: Seduced by Mathematics

> Euclid alone has looked on Beauty bare [1] — Edna St. Vincent Millay

"Seduced" is one of those words that now is used to convey a meaning that's the opposite of what was originally intended.

In my late teens, I was dining with the family of a friend of mine. My friend, my friend's parents, and their 7-year-old daughter were gathered around the table. All of a sudden, from out of nowhere, the daughter asked, "Daddy, what does 'seduce' mean?"

You could have heard a pin drop.

Daddy, however, was more than up to the challenge. Without missing a beat, he said, "The word 'seduce' comes from the Latin *duco*, meaning 'to lead', and *se*, meaning 'away from'. And right now you're eating your vegetables like a good girl, but I'll bet if I put a bowl of ice cream on the table, the ice cream would seduce you from eating your vegetables."

As good a real-time save as I've ever heard.

But nowadays, "seduce" is a good thing — or at the very least, an attractive thing. I just googled "seductive blend," and found 42,000 hits, starting off with a villa that is a seductive blend of style and serenity, and a cocktail that is a seductive blend of vodka and pomegranate juice.

I'm guessing that the title of this book, *Seduced by Mathematics*, will elicit one of two reactions. There will be a large number of people who will know *exactly* what I'm talking about, and (unfortunately) a larger number of people whose reaction will be "Are you kidding?" But I'm writing this book for both groups — to enchant the former and persuade the latter. Just as seduction can be accompanied (or accomplished) by a simple flower or an elegant bouquet, there are simple and elegant examples of the seductive in mathematics. And here's one, with a little background.

Math is a universal language — one which unlocks an amazing and incredibly useful collection of truths. But — as Edna St. Vincent Millay so eloquently wrote — it has an aesthetic that, unfortunately, most people who are exposed to mathematics don't appreciate.

I'm hoping I can contribute to changing that.

Seduced by Murder Mysteries

Many murders are solved through obvious clues, such as videos or fingerprints. And none of these murders make for a good murder mystery. What makes a murder mystery seductive is a detective who, through various lines of reasoning, deduces that the murderer must have been a left-handed dentist who recently visited Sweden.

And this is similar to mathematics. The vast majority of mathematics problems are solved via known formulas — and nobody considers this seductive. Formulas are the aphorisms of mathematics; they crystallize truth — but like aphorisms, some are definitely more appealing than others. My favorite aphorism, which I've NEVER had a chance to put into practice, distinguishes between coral snakes and king snakes. The bite of a coral snake can be fatal, not so for king snakes — but the two look much alike. The aphorism in the form of a rhyme distinguishes between the appearance of the two. "Red touch yellow, kill a fellow, red touch black, safe for Jack" [2].

Yes, there are some formulas that crystallize such an important truth that we memorize them, in part because we use them so frequently. The Pythagorean Theorem — which will make an appearance shortly — is one of those. But sometimes there's a captivating line of reasoning that underlies the formulas — and that's when math becomes seductive. So let me try to illustrate with an example that I recently encountered.

The World Series (and Other Best of ... Competitions)

The championship of professional baseball in the United States is called the World Series, and consists of a best-of-seven playoff between the winners of the American and National Leagues. Many other American sports feature this best-of-seven format for the playoffs — but a volleyball or table tennis match has a best-of-five format, and there are doubtless other competitions with similar methods to determine the winner.

A friend of mine recently decided to examine the probability that a World Series between two evenly-matched teams (teams which had an equal chance to win any game) would last a given length. In the case of the World Series, it will either last 4, 5, 6, or 7 games. There's a formula for this (you don't need to know it), and he was somewhat surprised when he discovered that the probability of the World Series lasting 7 games was the same as the probability of the World Series lasting 6 games. He asked me if there was an underlying reason for this.

I gave it some thought, and then I had it. Suppose that, after 5 games, the Series is still alive — one team has won 3 games, and the other has won 2. If the team that won 3 games wins game 6, the Series is over in 6 games. If the team that won 2 games wins game 6, the Series is over in 7 games. Since there is an equal chance of either team winning game 6, there is an equal chance that the Series will last 6 games or 7 games.

And that's an example of one of the things that those of us who have been seduced by math find seductive. It's not using the formula to find that the probability of a Series lasting 6 games is the same as the probability of a Series lasting 7 games, it's finding an underlying reason why this is so — and then realizing that it applies to ANY best-of … contests between two evenly matched teams. If, somewhere, there is a best-of-37 contest between two evenly-matched teams and the winner is the first contestant to win 19 games, the probability that the contest will last 36 games is the same as the probability that the contest will last 37 games.

A Seductive Proof of the Pythagorean Theorem

In the autumn of 1955, I was a sophomore in high school. I'd always been good at math, breezing through arithmetic and algebra — and then I hit geometry. Geometry hit back — hard. The text we used was a classic of the period; *Plane Geometry*, by Welchons and Krickenberger [3]. It presented the material in a terse style, with postulates and theorems, the latter being presented in two-column proof form. It required an effort on my part simply to get Bs, something that had never happened to me in a math course before.

Presenting mathematical proofs was something that was relatively new to me, as it is to most geometry students. It wasn't enough to just perform calculations or solve equations like I did in arithmetic and algebra, I had to create logical arguments and present them step-by-step. I was dinged when I left out steps that struck me as obvious. Algebra was fun, geometry was painful. And even though I hadn't yet come across Millay's poem, I sure didn't see any beauty in geometry.

After a few months, we arrived at what the teacher assured us was one of the high points of the course — the Pythagorean Theorem. The teacher told us that this result was so profound that, when Pythagoras had proved it, he ordered a celebration at which one hundred oxen were barbecued.

Below is the same proof of the Pythagorean Theorem as the one to which I was exposed in Welchons and Krickenberger, in a slightly easier-to-read format and with all the details filled in (the Welchons and Krickenberger proof had the annoying habit of sticking in a lot of "Why?"s for the student to fill in).

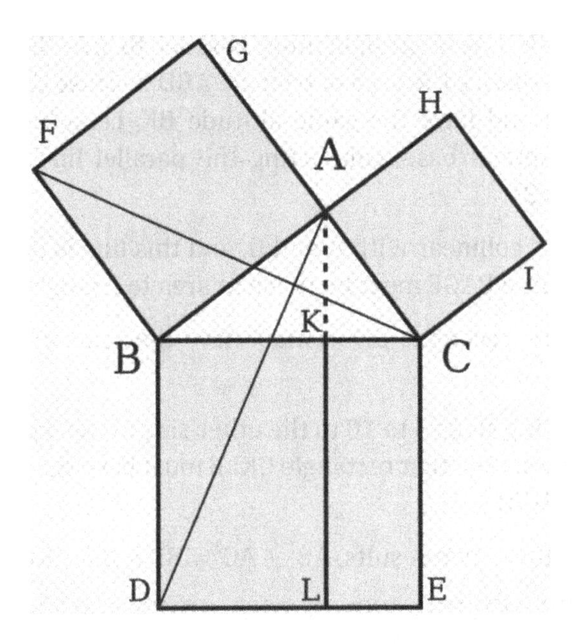

(1) Let ACB be a right-angled triangle with right angle CAB.

(2) On each of the sides BC, AB, and CA, squares are drawn, CBDE, BAGF, and ACIH, in that order. The construction of squares requires the immediately preceding theorems in Euclid, and depends upon the parallel postulate.[9]

(3) From A, draw a line parallel to BD and CE. It will perpendicularly intersect BC and DE at K and L, respectively.

(4) Join CF and AD, to form the triangles BCF and BDA.

(5) Angles CAB and BAG are both right angles; therefore C, A, and G are *collinear*.

(6) Angles CBD and FBA are both right angles; therefore angle ABD equals angle FBC, since both are the sum of a right angle and angle ABC.

(7) Since AB is equal to FB, BD is equal to BC and angle ABD equals angle FBC, triangle ABD must be congruent to triangle FBC.

(8) Since A–K–L is a straight line, parallel to BD, then rectangle BDLK has twice the area of triangle ABD because they share the base BD and have the same altitude BK, i.e. a line normal to their common base, connecting the parallel lines BD and AL (Lemma 2).

(9) Since C is collinear with A and G, and this line is parallel to FB, then square BAGF must be twice in area to triangle FBC.

(10) Therefore, rectangle BDLK must have the same area as square $BAGF = AB^2$.

(11) By applying steps 3 to 10 to the other side of the figure, it can be similarly shown that rectangle CKLE must have the same area as square $ACIH = AC^2$.

(12) Adding these two results, $AB^2 + AC^2 = BD \times BK + KL \times KC$

(13) Since $BD = KL$, $BD \times BK + KL \times KC = BD(BK + KC) = BD \times BC$

(14) Therefore, $AB^2 + AC^2 = BC^2$, since CBDE is a square [4].

Ok, so I knew enough about geometry to know that others were more capable than I of judging how important something was, but this barely seemed worth frying up a hamburger, to say nothing of barbecuing a hundred oxen. And had Edna encountered this presentation of the proof of the Pythagorean Theorem, we might have been short one poem from the collected works of Edna St. Vincent Millay.

I think I ended up with a B in geometry, but the next year was Algebra II, and mathematics was enjoyable again. A few years passed, and — almost certainly to the surprise of my geometry teacher, had she been aware of it — I became a math teacher myself. And one day I happened upon the following proof for the same Pythagorean Theorem (maybe he had more than one theorem, but this is the one we all know).

It starts with the following diagram, in which each side of the large square is the sum of the lengths of the sides of the triangle other than the hypotenuse.

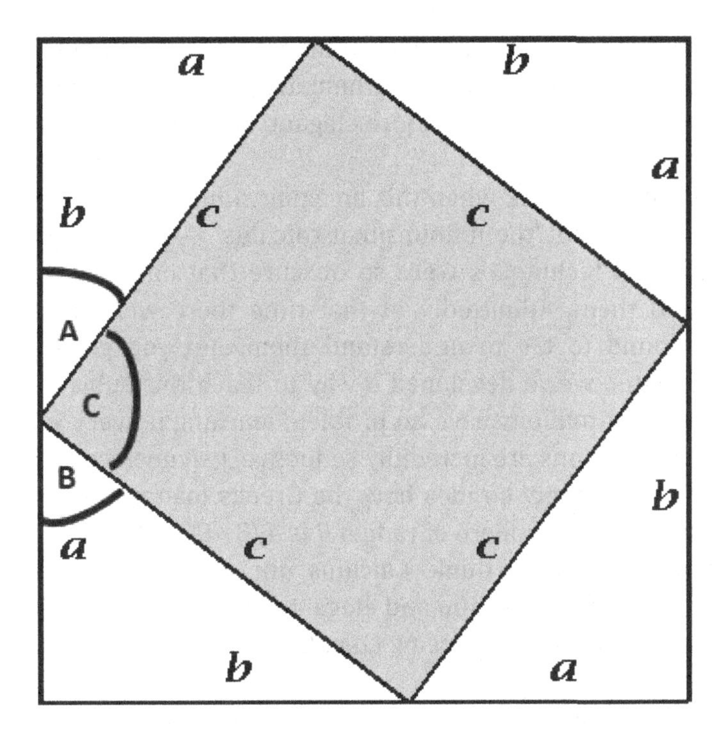

The four-sided gray object in the center sure looks like a square — but we've got to prove it is because pictures can be deceptive. Let C be one of the four angles of the gray object. If we let A denote the angle opposite side of the triangle, and B the angle opposite side b, then A + B + C is a straight angle of 180°. But in each one of the right triangles, A + B + 90° = 180°, since the sum of the angles in a triangle is 180°. So A + B = 90°, and therefore C = 90°.

The area of the large square is $(a + b)^2 = a^2 + 2ab + b^2$, and since the area of the large square is the sum of the inside square plus four right triangles, we see that $a^2 + 2ab + b^2 = c^2 + 4(½\ ab) = c^2 + 2ab$. Subtracting, we see that $a^2 + b^2 = c^2$.

The first time almost anyone sees this proof, the reaction is, "OMG, that's cool." And if it wasn't your reaction, this may not be the book for you. Granted, Pythagoras probably would have had to know some algebra to come up with that proof, although there's a simple geometric proof of the only algebraic fact used, that $(a + b)^2 = a^2 + 2ab + b^2$. And very rarely is the first proof of a theorem — or the first solution of a problem — the most beautiful, or the most insightful. But over the years, the combined efforts of mathematicians have found solutions to problems that are simpler, more elegant, and more insightful — in a word, seductive.

Take calculus. Back when this amazing subject was being developed, it was called "the infinitesimal calculus" — and its rules and computational techniques were so obscure that only a few people understood them. Admittedly, at that time there were only a few people around to try to understand them, but you get the idea. Centuries later, we've developed a way to teach this subject so that literally tens of millions have no problem learning it every year. And some of its solutions are incredibly seductive. I shamefacedly confess that I have absolutely no idea how the Greeks managed to compute that the volume of a sphere of radius R is $4/3\ \pi R^3$ — but with the aid of calculus it's a slam dunk. Calculus unravels so many of these Gordian knots in so beautiful and elegant a fashion that I could write an entire book called *Seduced by Calculus* — but I've confined it to a chapter or so here.

So that's what this book is — my shot at showing you — using a collection of some of what I believe are elegant and intriguing examples of interesting problems and fascinating concepts — that you *can* be seduced by math.

Bibliography

[1] Poetry Foundation. Online at https://www.poetryfoundation.org/poems/148566/euclid-alone-has-looked-on-beauty-bare.

[2] Madison Audobon. Online at https://madisonaudubon.org/blog/2020/6/12/red-touch-yellow.

[3] A. M. Welchons and W. R. Krickenberger, *Plane Geometry*, Boston, MA: Ginn and Company, 1949.

[4] Wikipedia. Online at https://en.wikipedia.org/wiki/Pythagorean_theorem Both of the proofs presented in this chapter can be found at this reference.

Chapter 1

Seduced by Numbers

I can remember the first fascinating thing I learned about numbers — they go on and on and on, without stopping.

OK, maybe not so fascinating to you the reader, but I was about 5 or 6 years old when my parents sent me to a summer camp, and I found a like-minded colleague. Maybe it's a bit of a stretch to call another 5- or 6-year-old boy a colleague, but during that summer at summer camp we engaged in the Continued Counting Project (CCP). I can't remember who started it, but one of us said: "One", the other responded "Two", the first individual said "Three" — and we continued throughout the summer. I can remember that when we were resting in the swimming pool, supported by our elbows on the side of the pool, we made sure that we were next to each other so that the CCP could continue. Likewise during the obligatory rest periods, when everyone else was napping we lay next to each other and continued the count *sotto voce* so that counselors couldn't hear us. I've talked to a number of other mathematicians who can recount similar experiences.

But they're mathematicians, you might say, they're pre-disposed to think that way and be attracted to numbers. But how likely is it that two young boys, out of a cadre of maybe 20 kids at summer camp, would find each other and engage in such a project?

Not as unlikely as you might think. According to a 2019 survey [1] (which repeats information I first heard in the 1960s, and then again in the 1990s), math and science are the favorite subjects of elementary school students. By the time they reach middle school, however — not so much. So somewhere along the line, parents and teachers are killing kids' natural enthusiasm for these subjects — and so I want to start off this book with what may be the best anecdote I've ever heard about the importance of numbers.

Size Matters

The world of mathematics and math education owes a great debt to Jordan Ellenberg, author of *How Not To Be Wrong* [2]. Jordan dug this anecdote up from who knows where and put it first in his book. And I owe a great debt to Jordan Ellenberg, as I was fortunate enough to interview him on his book — and after a scintillating 1-hour interview I realized I had forgot to turn on the recorder! Jordan kindly consented to a retake, and the second interview was every bit as good as the first.

So the year is 1943. Somewhere in the hills of New Mexico, the top physicists and chemists of the Allies are working feverishly on an effort, code-named the Manhattan Project, to develop the first atomic bomb. Of course, you know this, it's been the subject of countless books, movies, and TV shows. But at the same time, the left-over mathematics experts — mostly statisticians — were actually in Manhattan — to be precise, at 401 West 118th Street, at Columbia University. The Statistical Research Group (SRG) was engaged in solving mathematically-related problems for the military. Ellenberg's description of the intellectual firepower comprising the SRG is neatly summarized by saying that Milton Friedman, a future Nobel-Prize winning economist, was generally the fourth-brightest person in the room.

Let's turn our attention to Abraham Wald, the brightest person in the room. Wald had been born in Austria, but in the mid-1930s Austria was no place for someone named Wald. Fortunately, Wald had come to America at the behest of Oskar Morgenstern, co-author with the legendary John von Neumann of the seminal tract on game theory, and was currently a statistics professor at Columbia when not working with the SRG. On this day, the Army was confronted with a problem — where to put armor on fighter planes so that they would be less likely to be shot down? Of course, you couldn't put armor on the whole plane, it would be too heavy to fly.

The Army had compiled the following statistics on returning planes by counting the number of bullet holes in various sections of the plane. Wald — and the others in SRG — were confronted with the following data.

Section of Plane	Bullet Holes per Square Foot
Engine	1.11
Fuselage	1.73
Fuel System	1.55
Rest of the Plane	1.8

When the Army approached the SRG, they were predisposed to put the armor on the parts of the plane that were getting hit the most. But Wald quickly saw that they were looking at the problem incorrectly.

What Wald realized is that the numbers in the table showed vulnerability — the fewer bullet holes in a section, the more vulnerable the plane was in that area, and more likely to be shot down. But the words in the previous sentence may not convince you. To see that this answer is correct, let's look at an extreme situation. Ask yourself what it would mean if the number in the Engine line were 0, rather than 1.11.

No, it wouldn't mean that no bullets ever hit the engine. It would mean that EVERY time a bullet hit the engine, the plane failed to return. The numbers are a measure of vulnerability, with smaller numbers representing greater vulnerability. You can also see this by imagining that if one number were considerably higher than the others; it would mean that bullet hits in that section were less likely to bring the plane down.

This example goes to the heart and soul of what is seductive about math. It is so puzzling that the Army military needs to seek the help of the brightest mathematicians to solve it — and the solution can be understood by elementary school students.

Prime Numbers

Almost every important complicated object is built out of simpler ones. Organisms are constructed of cells. Molecules are assemblages of atoms. Protons are combinations of quarks.

And the positive integers are constructed by using products of prime numbers. Moreover, thanks to the Fundamental Theorem of Arithmetic, every number can be uniquely written up to the order of the factors as a product of primes. 90 is the product of 2, two 3s, and 5, and that's the only way to do it, as long as we are willing to accept that $3 \times 3 \times 2 \times 5$ is the same "prime decomposition" as $5 \times 3 \times 2 \times 3$.

Prime numbers are generally defined as "integers bigger than 1 which have no integer factors other than that integer and 1". Examples are given showing 5 is prime and 12 isn't. But for some reason, there's an awful lot of interest in the question "Why isn't 1 a prime? After all, the only integer factor of 1 is 1, so doesn't it fulfill the conditions? Why make an exception for 1?"

It's just a shame we can't talk about the Fundamental Theorem of Arithmetic before we give a formal definition of a prime number, because if we allowed 1 to be a prime number, the Fundamental Theorem of Arithmetic loses some of its beauty. We need to modify the "unique prime decomposition of a number" by allowing for the possibility of an arbitrary number of 1s. Sometimes you have a result in which you can't avoid something like "this theorem is true except for numbers with a remainder of 3 when divided by 7". But when you can, you do.

Anyway, the first question about primes is "How many prime numbers are there?" This question was answered by Euclid, with a beautiful proof of why there an infinite number of primes. Suppose, he said, that there were only a finite number of primes. Multiply all these primes together, and add 1. Call that number N; it's certainly bigger than each of the primes. If you divide N by any one of the primes, you get a remainder of 1, and so N must be prime, which contradicts the assumption that we made. For instance, if the only primes were 2, 3, and 5, then N would be $2 \times 3 \times 5 + 1 = 61$, and 61 has a remainder of 1 when divided by either 2, 3, or 5.

One of the first things that people noticed about prime numbers is that the distance between successive prime numbers seem to lengthen as the prime numbers got larger. Of course, this wasn't true in every case (more about that in a moment), but in a sense that could be quantified. It's sort of like the idea that the further you get from St. Louis, the harder it is to find fans of the St, Louis Cardinals. Oh, every so often you might stumble on a cluster of them, but you get the idea.

Let's go back to the prime numbers. All primes bigger than 2 must be odd, but every so often there are two odd numbers next to each other, both of which are primes — such as the pair 17 and 19, or the pair 41 and 43. Such pairs of primes are called twin primes. At the moment, no modern-day Euclid has an answer to the question of whether there are an infinite number of pairs of twin primes.

In 2013, academics who had taken their best shot at the Twin Prime Conjecture and not gotten very far were stunned to discover that Yitan Zhang, a little-known mathematician who was working at the time as the manager of a Subway franchise, had taken a huge stride towards solving this problem. There is an easy-to-understand term called "gap" in this area of mathematics, the gap between two primes is simply their distance. Using this term, the Twin Prime Conjecture can be stated as "there are an infinite number of prime pairs separated by a gap of 2". What Zhang showed was that "there are an infinite number of prime pairs separated by a gap of at most 600,000". In the years since the size of the gap in Zhang's result has been reduced to approximately 200 by extending some of the novel arguments that Zhang used [3].

There's an even more famous conjecture concerning primes than the one concerning twin primes which has a similar history. The Goldbach Conjecture, originally proposed in 1742, is really simple to state — it says that every even number is the sum of two primes. In 1939, Schnirelman proved that every even number is the sum of not more than 300,000 primes [4]. They've been chipping away at the number 300,000 in the 80 years since — but the Goldbach Conjecture, like the Twin Primes Conjecture, is still unproven.

I don't think it's possible for a relative outsider like Zhang to make a similar impact in the natural sciences. Lots of the big developments are made by large multi-organization teams or small groups mostly at a single institution. Plus, cutting-edge scientific equipment is expensive. But in Zhang you have an individual who worked on this one problem for a decade. And that's an excellent example of how math can seduce you. You find a problem, and you simply can't let it go.

Some of the problems, like the Twin Prime Conjecture, seem to require a serious mathematical background (which Zhang had) in order to make progress. But not necessarily, and in this book you'll see a number of problems which are easy to comprehend and still unsolved. My hope is that readers might be tempted to dig in. There's one in particular, which you'll see in the chapter on geometry, which is still unsolved and might actually be solved by a 6-year-old — if that 6-year-old gets lucky.

Seductive Numbers

Over the years, there have been several numbers which have attracted considerable attention. Yes, some of the integers fall in this category, mostly because they are thought to have some sort of extra-mathematical significance.

Three is one such integer. There are some intriguing and fairly comprehensible theorems in which three plays a stellar role. For instance, Gauss' Eureka Theorem, proved when he was 19 years old, is that every integer is the sum of three triangular numbers. We'll see triangular numbers later in this book — what we're concentrating on here is the number three. Gauss' Eureka Theorem is not a theorem about three — it's a theorem about triangular numbers.

What we're interested in is numbers that are seductive by themselves. And the first candidate is

A secret brotherhood of elites. A discovery which would shatter their most cherished beliefs. A member of the brotherhood who broke the most sacred rules of the brotherhood by divulging that discovery — and for that he was murdered.

A story ripped from today's headlines? Possibly — but I'm not referring to QAnon or contemporary conspiracy theories. The brotherhood was the Pythagorean Brotherhood, headed by (not surprisingly) Pythagoras, who not only had a reputation as a mathematician, but also a philosopher.

Among the Pythagorean Brotherhood, one of the fundamental beliefs was that the Laws of the Universe were governed by whole numbers. Yes, there were important numbers such as 1/2 or 3/5, but these fractions were quotients of whole numbers. The Greeks thought of 3/5 as the ratio 3 to 5, and so numbers such as 3/5 are known as rational numbers. In fact, the motto of the Pythagorean Brotherhood was "All is number". And number for them meant whole numbers, or ratios of them.

Hippasus of Metapontum [5] was a member of the brotherhood who discovered the secret that would shatter this belief — the square root of 2, which was the length of the hypotenuse of a right triangle with two sides each of length 1, was not such a ratio. As this could be demonstrated mathematically, this was known to the Brotherhood — but it was felt that if it were widely known the influence of the Brotherhood would be greatly diminished, as All was Not Number.

Hippasus, however, made the mistake of communicating this result to the *hoi polloi* — those not members of the elite. And, on a sea voyage with some members of the Brotherhood, he was thrown overboard. At least, so the story goes — and you probably wouldn't be surprised to learn that similar tales of secret societies, shattering discoveries, and consequent murders have happened in many other situations in the 2,500 years since these events are presumed to have occurred.

Hippasus' original proof was geometric in nature, but nowadays it's generally proved by one of the most important tools in the mathematical toolbox — proof by contradiction. The idea is to show a statement is true by assuming it is false, and showing that this assumption leads to a contradiction. As long as the statement we are investigating is either true or false, once you have eliminated the possibility that the statement is false, you must accept the fact that the statement is true.

So here goes. Assume that the square root of 2 can be written as a quotient p/q of whole numbers, where p and q have no common prime factors. Of course, reducing a fraction to its simplest form by canceling common factors in the numerator and denominator is something you saw in elementary school. So

$$\frac{p}{q} = \sqrt{2}$$

Squaring both sides of that equation and then multiplying both sides of the equation by q^2 results in the equation $p^2 = 2q^2$. $2q^2$ is an even number, which means that p^2 must be even. Since the square of an odd number is odd, p cannot be odd, and so p must be even. Consequently $p = 2k$ for some integer k.

But now, if we substitute $2k$ for p, we get $4k^2 = (2k)^2 = p^2 = 2q^2$. Dividing the leftmost and rightmost terms by the common factor 2, we get $2k^2 = q^2$. But $2k^2$ is even, and so the exact same reasoning that we just employed show that q must be even, and so $q = 2j$ for some integer. But there's the contradiction, because we initially chose p and q so that they had no common prime factor — and here we just established that 2 is a common factor of both p and q. Therefore the assumption that the square root of 2 is the ratio of whole numbers must be false.

You might think that there's something absurd about basing a belief structure on whole numbers — but after all, it was 2,500 years ago, and we're much more sophisticated now. Or are we? We can hear an echo of the Pythagorean Brotherhood in the statement of the 19th century German mathematician Leopold Kronecker, who declared that "God made the integers, all else is the work of Man" [6].

But mathematicians are nothing if not dedicated structure-seekers, and it soon proved possible to construct a mathematical structure which contained the ratios of whole numbers, the square root of 2 — and a whole lot more.

Algebraic Numbers

The fraction 3/5 is a solution of the equation $5x - 3 = 0$; in other words, it is the root of a polynomial of degree 1 with integer coefficients. The square root of 2 is a solution of the equation $x^2 - 2 = 0$, a polynomial of degree 2 with integer coefficients. An algebraic number is any real number that is the root of a polynomial of any degree with integer coefficients.

There are a LOT of algebraic numbers. In fact, the sum of two algebraic numbers is an algebraic number, the product of two algebraic numbers is an algebraic number, and the reciprocal of an algebraic number is an algebraic number. There is an interesting proof of this due to the German mathematician Hecke, but it's a little beyond the scope of this book. So the question arises — is that all there is? If there are numbers that are not rational, could there be numbers which were not algebraic?

π

The answer to the last question is yes — there are numbers which are not algebraic. And at least one of them, π, was known to the ancients. In fact, if you want to build a religion based on a number, π might be a really good choice, if for no other reason than that π has a great publicity department. Is there any other number which has inspired books and movies?

π is such an important number that it will continually reappear throughout this book, but let me give a couple of trailers for π that will demonstrate its versatility. The first is that π exerts a tremendous influence when it does not appear — or, at least, appears behind the scenes. This happens when we measure angles in radians.

Degree measurements have a long and honored history, but they're an idiosyncrasy generated by the fact that the Earth revolves around the Sun in 365 days. 365 is a horrible number to work with, but 360 is great — it has lots of divisors, enabling the circle to be split up easily into sectors. But doing so creates a horribly ugly fudge factor of $\pi/180$ in a number of crucial formulas.

The formula for the arc length s of a circle of radius r subtended by a central angle ϑ is a ghastly

$$s = \pi/180\ r\vartheta$$

if we measure the angle ϑ in degrees. Who wants to memorize — and use — an unwieldy formula like that when we have the exquisitely simple

$$s = r\vartheta$$

available if we measure ϑ in radians, of which there are 2π in a full circle. It also impacts the formula for the area of a circular sector similarly, giving

$$A = \tfrac{1}{2}\, r^2\, \vartheta$$

when the central angle is measured in radians. It's a lot nastier if the coefficient is $\pi/360$ rather than ½, as it would be if the angle were measured in degrees.

When we look at calculus later on in this book, we'll see another instance of how valuable radian measure is.

Ok, now on to the sublime. The digits of π never repeat in the "forever" sense, of course — else π would be a rational number. So you're going to see two 3s together somewhere down the line if you start looking at the digits of π — actually, you'll see them at digits 24 and 25. You'll also find the number 1,729 (good story about this number later on) somewhere in π. In fact, you can find any finite string of digits somewhere in π.

And therein lies something truly intriguing. The coding system ASCII (American Standard Code for Information Interchange) translates every symbol used in printing this book into a five-digit integer. Doing so by writing down the five-digit ASCII integer representing the first printed character in the book, then following that with the five-digit ASCII integer representing the second printed character in the book, etc. will result in an integer that is at most a million digits long — and we can find that very long integer somewhere down the line in the digits of π.

So in a sense this book has already been written, and if you look hard enough, you can find it in the digits of π. But not just this book lies in the digits of π. Every book ever written — or every book that ever will be written, is contained in the digits of π. And although this was only first realized by von Lindemann, maybe Pythagoras was on to something when he said, "All is number".

e

The transcendental number *e* arises naturally in a discussion about compound interest. Suppose we invest \$1 at 100% interest for a year, at the end of the year the \$1 has doubled to \$2. This is computed as $(1 + 1)^1$.

Now let's suppose that instead of 100% interest for 1 year, we compound it semi-annually. The way this is done is to split the year into two half-years. Instead of one "settling up" at the end of the year, we split the year into two half-years, and "settle up" at the end of each half-year.

One hundred percent paid on a dollar for half a year results in interest of 50 cents, when we settle up at the end of the first half-year, we have \$1.50 in the account. One hundred percent paid on \$1.50 for half a year results in interest of 75 cents, when we settle up at the end of the year by adding 75 cents interest to the \$1.50, we obtain a total of \$2.25. We computed this as $(1 + 1/2)(1 + 1/2) = (1 + 1/2)^2$.

If we compounded quarterly (4 times annually), we would multiple 1 times $(1 + 1/4)$, then this times $(1 + 1/4)$, then that result times $(1 + 1/4)$ — and then that amount finally by $(1 + 1/4)$, which would give us a total amount of $(1 + 1/4)^4$ = \$2.44 (rounded to the nearest penny). If we compound N times a year, the amount in the account would be $(1 + 1/N)^N$. Compound monthly (12 times a year), and the amount is \$2.61. Compound daily (365 times a year) and the amount is \$2.71. We are making more by compounding more, but it seems that this is leveling off.

This brings us to one of the breakthrough moments in mathematics — the idea of what happens to a quantity that depends upon an input value as that input value gets larger and larger. Or smaller and smaller. Or closer and closer to something. There are two mammoth concepts here. The first is the quantity that depends upon an input value — now known as a function, written $y = f(x)$, where x is the input value (now called the independent variable) and y is the quantity that depends upon that input value, which is now called the dependent variable. The second is what happens to that function as the input value gets larger and larger, etc. — and that's called a limit.

In this case, the quantity $(1 + 1/N)^N$ does indeed get larger and larger as N itself gets larger and larger, and it gets as close as possible to $e = 2.71828 \ldots$ without ever getting there. It isn't easy to show this without knowing some calculus, so I'm not going to bother and hope that those who aren't comfortable with calculus will just accept this, because mathematicians have proven it.

Like π, e is a transcendental number. However, the fundamental role that it plays is not so evident as that played by π. We get familiar with circles very early in life and education. Exponential functions, not so much — and that's where e comes to the fore. After all, most of us don't acquire a bank account until high school at the earliest, and even if we leave money in it to compound, the compounding process need not involve the number e. When we take a high school algebra course, we are taught about the exponential functions $y = 2^x$, because it is tied to the doubling period of a growing quantity, and the function $y = (1/2)^x$, because it is tied to the half-life of a decaying quantity. But we are force-fed $y = e^x$ with only the assurance that later on, we'll fully appreciate it, but learn it for now.

This really seems a little silly, because the function $y = 10^{kx}$ describes all the exponential functions as long as we choose k appropriately. And because there's a log button on our calculators, which is $\log_{10}$, we soon learn that $a^x = 10^{(\log a)x}$, so why everyone is making such a fuss about e is unclear.

And it's not clear if you go to the Internet, either — try searching "Why are natural logarithms natural?" and see what you get. There might be some good stuff lurking further down the line than the first few pages, but I didn't see it. The best explanation I can give requires some calculus to understand it thoroughly, but I can give an intuitive explanation.

Most of us have turned on a water faucet at some time. During that short period where we are turning the faucet, the rate of liquid flow is changing — it starts with a trickle and increases to a vigorous flow. If we were to stop turning the faucet at any moment, the rate of water flow would be constant. That rate of flow — when we stop turning the faucet (even if we haven't turned it as far as it will go) — is called the instantaneous rate of change (in this case, of the volume coming from the faucet).

Exponential functions are qualitatively like turning the faucet, the instantaneous rate of change increases as time increases. But of all the exponential functions, $y = e^x$ is the ONLY one for which the instantaneous rate of change of the volume is the same as the volume at that moment. If we let $y = e^x$ denote the amount of water at time x (remember, we are REQUIRING the amount of water to satisfy this condition), at any time the instantaneous rate of change of the volume is the same as the actual volume. Not twice the volume, not half the volume — but exactly the volume.

The statement that a growing or decaying quantity is proportional to the amount of that quantity is a natural one to make. If there are twice as many rabbits in one place as another, it is reasonable to assume that twice as many rabbits will create twice as many bunnies. That's growth proportional to the existing quantity. Exponential functions describe these quantities — and the number e has the property that it is the only base for an exponential function for which the instantaneous rate of growth of a quantity is precisely the same number as the actual quantity.

In a sense, e is "natural" for growth functions in the same way that radians are "natural" for angle measurements.

Reciprocity

I can't resist closing out this section with a mention of an interesting parallel between π and e. Their reciprocals are both solutions to intriguing problems in probability.

The Buffon Needle Problem is the oldest problem in the field of geometrical probability. Imagine that we have a needle of length 1, and we also have a large piece of paper with a large number of parallel lines 1 unit apart. The needle is dropped randomly on the piece of paper. The probability that the needle will cross one of the lines is $1/\pi$.

I'm not sure what the next problem is called, but I think of it as the Marriage Problem, because the problem is very interesting in the context of how one decides on a strategy for choosing a mate. But I'm going to set it up in a different context because it's easier to explain it initially, and then show how it applies to marriage.

Let's suppose you have a number of envelopes in front of you, each containing a different amount of money. You can open any envelope and look at the amount of money in it — but you must then decide whether to take the money, or choose another envelope. Once you take the money, the game is over — and you can never go back and choose an envelope you previously opened.

A reasonable strategy is to decide to open some fraction of the envelopes and keep track of the largest amount of money in the opened envelopes. Think of this as the comparison pool. Then, start opening the other envelopes. As soon as you find one with more money than the largest amount in the comparison pool, take it — or take the money in the last envelope if push comes to shove. So the question is — what fraction of the available envelopes should you use as the comparison pool? It turns out that you should multiply the number of envelopes by $1/e$, and use that as the comparison pool.

So how does that apply to the Marriage Problem? Most people have a limited time during which they'll contemplate marriage — for purposes of illustration, let's say it's from 18 to 40. That's a total of 22 years. Multiply 22 by $1/e$, that's about 8.1 years. Use the years from 18 to 26.1 to form your comparison pool, and then marry (or try to marry) the first person whom you judge superior to the choices in your comparison pool.

But don't blame me — or e — if this doesn't work for you. Like any probabilistic venture, the best strategy doesn't always work, but it works best in the long run — which in this case is for most people. I sometimes wonder if we don't actually do something like this subconsciously.

Bibliography

[1] Education Quizzes. Online at https://www.educationquizzes.com/surveys/what-is-the-favourite-school-subject/.
[2] J. Ellenberg, *How Not To Be Wrong*. New York, NY: Penguin Press, 2014.
[3] V. Neale, *Closing the Gap*. Oxford, UK: Oxford University Press, 2017.
[4] L. G. Schnirelman, *On the Additive Properties of Numbers. Proceedings of the Don Polytechnic Institute in Novocherkassk*, Russia, 1930.
[5] Wikipedia. Online at https://en.wikipedia.org/wiki/Hippasus.
[6] Wikipedia. Online at https://en.wikipedia.org/wiki/Leopold_Kronecker.

Chapter 2

Seduced by Arithmetic

Arithmetic is where numbers get together to produce other numbers. They do so by means of the four basic arithmetic operations: addition, subtraction, multiplication, and division. There are simple models for each of these — addition is continued counting, subtraction is take-away, multiplication is repeated addition of the same quantity, and division is sharing fairly. There's another model for division — successive subtraction — which is analogous to the repeated addition model for multiplication, but is not as intuitive or as useful as the sharing model.

Last Digit Standing

Art Benjamin devised a fascinating demonstration I call "Last Digit Standing" [1], which my wife Linda and I have enjoyed several times when he does it at *The Magic Castle* in Hollywood. I've revised it a little to do a slightly different presentation, which I've used just before the students do teacher evaluations. It amazes my students, which leads me to believe that we're seriously deficient in teaching arithmetic these days, preferring to rely on having them learn how to use a calculator.

One of the reasons that I'm writing this book is to be a resource for teachers. Math can be entertaining, and an entertaining teacher is usually a good teacher, if for no other reason than they've taken the effort to go beyond the subject matter and look at how to make students want to absorb the subject matter. As I look over my education, the teachers I remember best were generally the ones who entertained me.

The trick relies on the fact that the sum of the digits of a multiple of 9 is itself a multiple of 9 and vice versa — if the sum of the digits is a multiple of 9, then the number itself is a multiple of 9. This is easy to show — let's just see it for a three-digit number, which I'll write as $100H + 10T + U$, where H is the hundreds digit, T the tens digit, and U the units digit. The sum of the digits is $H + T + U$, and $100H + 10T + U = 99H + 9T + (H + T + U) = 9(11H + T) + (H + T + U)$. So if $H + T + U$ on the right is a multiple of 9, so is $100H + 10T + U$ on the left. If $100H + 10T + U$ is a multiple of 9, let's say it's 9X. Then $H + T + U = 9X–99H–9T$, which is clearly a multiple of 9.

So here's the trick as I perform it — and thanks, Art, because I get better teaching evaluations from my students when I do this. I remove the aces, tens and face cards from a deck of cards — leaving 32 cards. I then ask a student to blindfold me, after which I hold the 32-card deck face down, and ask ten different students to pick a card from the deck at random, and write the card number on the board with a times sign after it. After all ten have selected a card, I ask the students to use their calculators and write the product on the board. This might result in the board looking something like this

$$3 \times 4 \times 3 \times 9 \times 5 \times 8 \times 7 \times 6 \times 6 \times 7 = 22{,}861{,}440$$

I then ask another student to come up and circle one of the digits from 1 through 9 in the product — let's say the digit circled is 8. I ask the student to erase the remaining digits one by one in any order until only the circled digit remains — and call out each digit as it is erased. I then tell them what the circled digit is. What I do is simply add up the digits as they are called out — here's how it works. I've chosen a specific order for the erased digits, but the order doesn't matter — as you'll see shortly.

Erased Digit	Sum
2	2
4	6
0	6
4	10
1	11
2	13
6	19

At this stage, the student stops — and I tell them the last digit standing is 8. The product 22,861,440 is a multiple of 9, so the digits must add to a multiple of 9, and the next multiple of 9 after 19 is 27, so the remaining digit is 27 – 19 = 8. Of course, in order for the product to be a multiple of 9, the ten chosen cards must either contain a 9 or two 3s or two 6s or a 3 and a 6 — but that happens 39 times out of 40, so I've got really good odds, and it hasn't failed me yet.

1,729

G. H. Hardy [2] was one of the most brilliant mathematicians of the first half of the 20th century — but you probably wouldn't want him at a party, because not only was he dull, he knew he was dull — at least, to the outside world. To give you an example, he described an incident where he opened a letter from someone he had never met as the most romantic moment in his life.

The letter was from an unknown Indian clerk who had been working on his own on results in Hardy's specialty, number theory. Hardy described some of the results, which were unknown to him, as so beautiful they had to be true. Hardy invited the clerk, Srinivasan Ramanujan [3], to join him at Cambridge University.

The relationship between Hardy and Ramanujan deepened as Hardy's respect for Ramanujan's ability grew. However, the harsh winters and the English diet took a toll on Ramanujan, who fell ill.

Hardy visited Ramanujan at the hospital, and remarked to Ramanujan that the taxicab he rode over in was number 1,729, and that was such a dull number that he hoped it wasn't an omen. Ramanujan replied that 1,729 was not dull, it was fascinating, as not only could it be written in two different ways as the sum of two cubes, it was the smallest number with that property.

Ramanujan, off the top of his head, not only knew that $1{,}729 = 1{,}000 + 729 = 10^3 + 9^3$ and $1{,}729 = 1{,}728 + 1 = 12^3 + 1^3$, but that no number less than 1,729 could be written in such a fashion. Although many numbers could be written as the sum of two cubes (for instance, $9 = 8 + 1 = 2^3 + 1^3$), no number less than 1,729 can be written in two different ways as the sum of two cubes.

This is actually a little less amazing than it appears on first sight. I'd venture to say that the majority of people who are interested in math know the cubes of the numbers from 1 to 12 — even in this era when we have calculators that do all the number crunching. You'd only have to add up a few pairs of cubes to see this. A reasonable parallel might be for a soccer fan to know the four semifinalists in all the World Cups held this century.

But then there's Euler [4].

635,318,657

Nearly two centuries before Ramanujan, Leonhard Euler was unquestionably the greatest mathematician in an era that knew many groundbreaking mathematicians. While there is no evidence that Euler knew about 1,729, he knew about 635,318,657, which is expressible in two different ways as the sum of two fourth power. To be precise, $635{,}318{,}657 = 158^4 + 59^4 = 134^4 + 133^4$. And he knew this in an era without calculators! This is far more impressive, at least to me, than Ramanujan's knowledge of the peculiar property of 1,729. To know this, Euler would have had to have, at some stage, calculated all the fourth powers of the integers from 1 to about 160 (and done it correctly), added up the various pairs of fourth powers (and done that correctly), and made the critical observation. Continuing with the soccer analogy, maybe like knowing all the results of all matches from the quarterfinals on since the inception of the World Cup.

There is no evidence that Euler knew that 635,318,657 was the smallest number that could be expressed in two different ways as the sum of two fourth powers, but it was later shown that this was indeed so. Frankly, I'd be surprised if Euler knew this, but I'm surprised that Euler knew about 635,318,657 in the first place. What persuaded him to calculate all those fourth powers in the first place?

Fast forward to the year 1965. A friend of mine was studying number theory, and knew about both 1,729 and 635,318,657. There was to me an obvious question to ask, and so I asked it — what is the smallest such number which is expressible in two different ways as the sum of two fifth powers?

My friend researched the literature — not an easy job at the time, as there were no search engines with access to unreal amounts of data. He could find nothing, but he had access to a university computer, and burned up 20 hours of valuable (in 1965) computer time in a fruitless quest to find such a number. Remember, computers were rare, slow, and expensive in the 1960s. I'm not sure why his quest was terminated, but we found out later that, at the time, nothing was known about this problem. It wasn't even known if there was such a number.

And, as far as I know, it still isn't.

Taxicab Numbers

I must admit, I had a hard time deciding whether to put the stories of 1,729 and 635,318,657 in this chapter or in the chapter about numbers, but decided to put them here because there's a lot of arithmetic involved, especially exponentiation.

One of the ways to generate interesting mathematical conjectures is through the process of generalization. Ramanujan's number involves writing a number in TWO different ways as the sum of **two** *third* powers. Euler's number involves writing a number in TWO different ways as sum of **two** *fourth* powers. I've chosen three different ways to express numbers — CAPITALS, **bold** and *italic* to illustrate the three different variables in the taxicab numbers Taxicab(n,p,m), which is the smallest integer that can be written in m ways as the sum of n pth powers.

Using this notation, 1,729 = Taxicab(2,3,2), 635,318,657 = Taxicab(2,4,2), and, as I said at the conclusion of the last section, it isn't even known whether there is a Taxicab(2,5,2). But in 2002, Jaroslaw Wroblewski and Stuart Gascoigne proved that for any n, Taxicab(n, $n+1$, $n+1$) exists — in other words, for any n there is a number which can be written in $n+1$ different ways as the sum of n $(n+1)$st powers.

I'm including a proof in the Bibliography section, because I think it's ingenious. I've taken it verbatim from the Internet [5]; but I'm not presenting it here, as it may dissuade people from reading further, fearing that this proof is indicative of the difficulty level of what's to come. It isn't — most of what you'll see here isn't anywhere near this level, although the proof in the Bibliography section doesn't involve high-power mathematics.

There's an interesting point about this proof. It's non-constructive, and brings to mind a joke that was making the rounds when I was in grad school — and for all I know, made the rounds long before that.

An engineer, a physicist, and a mathematician were applying for a job with a company, and were required to take a test to demonstrate competence. They were shown to a room with a fire burning, a bucket, and a supply of water, and told to put out the fire.

The engineer went first. He grabbed, the bucket, filled it with water, poured it on the fire, and repeated the process until the fire was extinguished.

Next came the physicist. She calculated the volume of the bucket, the rate of growth of the fire, and poured precisely enough water on the fire to extinguish it without wasting any water.

Finally, the mathematician entered the room. He observed that enough water existed to put out the fire — and left.

The proof given in the Bibliography section is of that variety — it is known in the trade as an existence proof. It doesn't tell you ANYTHING about how to find any Taxicab number; it just tells you that there are certain types of them. So, if you were trying to find

Taxicab (4,5,5), at least you're not burning up time in a fruitless quest for something that may not even exist, like my friend's quest for Taxicab (2,5,2).

An Oldie But Goodie

There are some problems that were first proposed thousands of years ago. The one I'm about to cite appeared in a book on Arithmeticke (that's how they spelled it then) about the time of the Revolutionary War.

Problem: If a hen and a half lays an egg and a half in a day and a half, how many eggs will half a dozen hens lay in half a dozen days?

This problem is SO beautifully — and deceptively — stated, I'd like to meet the individual who created it, although it would require time travel to do so. The immediate picture that comes to mind is of one hen laying one egg in one day, and half a hen laying half an egg in half a day. You can just see the right half (or left half) of that hen laying half an egg in half a day. As a result of this line of thought, the temptation is to jump to the conclusion that one hen produces one egg in one day, so half a dozen hens will produce half a dozen eggs in one day, and half a dozen hens will produce three dozen eggs in half a dozen days.

Tempting — but wrong.

The difficulty is that in solving the problem this way, we've essentially thrown away some information. Let's work with ALL the information and see where it gets us.

A hen and a half lays an egg and a half in a day and a half, so half a dozen hens (four times a hen and a half) will lay six eggs (four times an egg and a half) in a day and a half. Since half a dozen days is four times as long as a day and a half, half a dozen hens will lay $24 = 4 \times 6$ eggs in half a dozen days.

There's another way to solve this using proportions from simple algebra. An obvious assumption is that each hen lays eggs at a certain rate, say k eggs per day. Therefore, the number of eggs E laid by H hens producing for D days is given by $E = kHD$. From the information given in the problem, $1\frac{1}{2} = k(1\frac{1}{2}) \times (1\frac{1}{2})$, so $k = 1/1\frac{1}{2} = 2/3$.

Therefore the number of eggs laid by 6 hens in 6 days is $2/3 \times 6 \times 6 = 24$, as shown above.

1,089

Yes, I know there are a lot of sections headed with numbers in this chapter, but it takes arithmetic to unearth their special beauty. 1,729 isn't just a number, it has a story associated with it courtesy of arithmetic.

And 1,089 does, too. Start with any three digit number where the first and last digits differ by more than 1. Let's use 927. Reverse the digits and subtract the smaller from the larger. If we reverse the digits, we get 729. Subtracting 729 from 927 gives 198. Now reverse the digits of this number, getting 891, and add it to the number whose digits we just reversed. Notice, that $198 + 891 = 1{,}089$. As long as you adhered to the rule that the original number had first and last digits differing by more than 1, you ALWAYS end up with 1,089.

My friend Al Posamentier uses this to interest youngsters in arithmetic. He works with a class of children, and asks each individual in the class to pick their own starting number (as long as it's one in which the first and last digits differ by more than 1), and go through the procedure. He then asks each child to read off their final number. They ALL end up with 1,089 — assuming they went through the process correctly — and the children can't wait to go home and show it to their parents.

I do similar demonstrations as Al — but I have a different collection of intriguing demos, I'll discuss a few of them at the appropriate juncture — I've already mentioned Last Digit Standing. But let's see why this trick works.

Let's assume we start with a number in which the hundreds digit is larger (by more than 1) than the units digit, and write it $100H + 10T + U$, as we did earlier in this chapter. If we reverse the digits, the number we create is $100U + 10T + H$ — since this is the smaller of the two, by assumption, we subtract it from the original number.

This gives us $(100H + 10T + U) - (100U + 10T + H) = 99H - 99U$. We need to know what this number is as a three-digit number, so we can figure out its hundreds, tens, and units digits. You may or may not be able to see immediately what those digits are — I certainly can't. But what I can do is try it out for a specific example and see if I can form a conjecture.

And I have a terrific example available — I already did it for 927 and got 198. It looks like the hundreds digit is 1 less than the difference between the original hundreds digits and tens digit. AHA! — that's why I have to have the difference between the original hundreds digit and the original tens digit to be greater than 1. Let me try another –256 and 652. Difference is 396. I think I've got it; the middle digit is always 9, and the sum of the hundreds and units digit must total 9.

If this were a SERIOUS book on math, I'd feel obligated to prove that conjecture — and it isn't really hard to do. But I'll try one more number, and if that's the way it is, I'll believe that it's always like that. 348 and 843 — difference should be 495, and indeed, it is.

So now we know we're adding $100h + 90 + (9–h)$ to $100(9–h) + 90 + h$. And what do we get? $900 + 90 + 90 + 9 = 1,089$.

And now you've got the basis for a magic trick using what magicians call a force — a method of controlling a choice made by a participant. In this case, since the result of this procedure is always 1,089, the magician can do something pre-planned with the number 1,089. The magician might take out a telephone book (if you can still find them), ask the participant to use the first three digits of the number he has obtained as the page number, and then use the last digit to locate the appropriate name on the page ("If the last digit is five, go to the fifth line on the page."). The magician now goes into a trance and says, "I'm seeing the name John Doe with phone number 342-1863." Of course, all the magician has to do is memorize the name and number of the ninth person on page 108, and if he or she can't do that, find another line of work.

Does the 1,089 Trick Work in Octal?

Human beings have ten fingers, a lot of counting is done using fingers, so it seems natural that our numbering system and the associated arithmetic is done using tens. We are familiar with the fact that the number 1,089 is 1 thousand (that's 1×10^3), 0 hundreds (that's 0×10^2), 8 tens (that's 8×10^1) and 9 units (that's 9×10^0). The digits that we use are 0, 1, 2, 3, 4, 5, 6, 7, 8, 9 – and there is no single digit for ten, because we use the idea of place value to write 10 – 1 is in the tens ($=10^1$) place.

However, it's very possible that inhabitants of other planets have only 8 fingers (or tentacles, or whatever) — and it would be natural for them to use only the digits 0, 1, 2, 3, 4, 5, 6, 7 and write 8 as 10, with 1 being in the eights ($=8^1$) place. Incidentally, most computer programmers have ten fingers, but they often use the octal number system because it has certain advantages when dealing with computers.

The 1,089 trick isn't going to work in exactly the same way in octal — partly because 8 and 9 aren't digits in octal. But it does work similarly, as we'll see. You have to know how to do octal addition and subtraction, but it's not hard. If you aren't already familiar with it, I'll explain it as I do it the first few times.

Let's start with a three digit octal number — 652. Let me first explain what this number is — using the idea of place value, it's $6 \times 8^2 + 5 \times 8^1 + 2 \times 8^0$. We could simply multiply this out and see how we would write it in decimal, but the inhabitants of our hypothetical planet — the ones with 8 fingers or tentacles — certainly wouldn't, so we're not going to, either. The first step is to reverse the digits, getting 256. Now we have to subtract the smaller number from the larger — and do it in octal.

$$\begin{array}{r} 6\ 5\ 2 \\ -2\ 5\ 6 \end{array}$$

If we were doing this in decimal, we can't subtract 6 from 2, so we'd need to borrow 1 from the tens column. I just HATE the word "borrow" in this context and prefer "exchange". We do the same thing in octal, we exchange a 1 from the eights place for eight units (or eight ones, whichever you prefer). We now have one less eight, and eight more units. That makes ten (speaking decimally) units, 2 in the original number and eight from the exchange. Our subtraction problem now looks like

$$\begin{array}{rrr} 6 & 4 & 10 \\ -2 & 5 & 6 \end{array}$$

We're good to go in the units place, but we can't take away 5 from 4 in the eights place, so we need to do another exchange. We exchange a 1 from the sixty-fours (= 8^2) place for eight eights, and the subtraction now looks like

$$\begin{array}{rrr} 5 & 12 & 10 \\ -2 & 5 & 6 \end{array}$$

And now we're cleared for takeoff. We can do the takeaway in each column, getting the answer 374 (remember, this is an octal number, representing 3 sixty-fours, 7 eights, and 4 ones). We now reverse the digits, getting 473, and do the octal addition. It's similar to octal subtraction, there are times we need to do exchanges if we have more of something than 7, because there is no single digit in octal representing a number more than 7.

$$\begin{array}{rrr} 3 & 7 & 4 \\ +4 & 7 & 3 \\ \hline 7 & 14 & 7 \end{array}$$

There is no single digit in octal for 14, so we exchange 8 of the eights (because it's in the eights column) for one sixty-four, and add that to the 7 in the sixty-fours column. This results in the octal number 867. But there is no single digit for 8 in octal, so we exchange those 8 sixty-fours for one item in the next column to the left (they are five-hundred-and-twelves = 8^3). That leaves us with 1 five-hundred-and-twelve, 0 sixty-fours, 6 eights, and 7 ones — 1,067 in octal.

Looks awfully like 1,089, doesn't it? If you think about it, it's essentially the same number relative to octal that 1,089 is in decimal, because 7 is the largest digit in octal (like 9 is in decimal), and 6 is the next largest digit in octal (like 8 is in decimal). And 0 and 1 are the smallest digits, respectively, in both systems.

6,174

I'm guessing that as long as there are people who are interested in numbers — and a lot of people, even non-mathematicians are — we will discover numbers with unexpected fascinating properties.

This next process first came to light in a paper by the Indian mathematician D. R. Kaprekar, written in 1955 [6]. This paper was titled "An Interesting Property of the Number 6,174." I'm also guessing that until this paper was written, the only people who were interested in 6,174 were people who lived at 6174 Elm Street, or a similar address.

So here's the interesting property. Start with any four-digit number which uses at least two different digits. If you decide to start with a number such as 137, which has only three digits, supply enough leading zeros to get a four-digit number — in this case, 0137. Then arrange the digits of the number in both ascending and descending order. Subtract the smaller from the larger, and repeat the process. Sooner or later, you'll get to 6,174. Always — as long as you did not start with a number such as 2,222, which has four identical digits. It actually won't take you that long, you'll get there in at most 8 repetitions of the basic process (arranging and subtracting).

Since 1,729 is such an intriguing number — as Hardy would now certainly agree — let's see what happens when we start with 1,729 and apply this process.

1,729	9,721 digits in descending order
	1,279 digits in ascending order
	8,442 subtracting the smaller number from the larger
8,442	8,442
	2,448
	5,994

5,994	9,954
	4,599
	5,355
5,355	5,553
	3,555
	1,998
1,998	9,981
	1,899
	8,082
8,082	8,820
	288
	8,532
8,532	8,532
	2,358
	6,174
6,174	7,641
	1,467
	6,174

You can see that whenever you get to 6,174, you've reached the end of the line — as applying the arranging and subtracting process (which is known as Kaprekar's routine) to 6,174 results in 6,174. 6,174 is called a fixed point of Kaprekar's routine — a fixed point of a process is one which the process doesn't change.

6,174 is known as Kaprekar's constant, as he was the person who discovered this remarkable property. The first question to be asked is — what on Earth enabled him to discover this? I'm tempted to ask whether he didn't he have anything better to do with his time — but I've spent a lot of time on problems that to the naked eye seem every bit as pointless. This is clearly an opportune moment to remember that injunction about people who live in glass houses not throwing stones.

I spent the second half of my research career investigating fixed points. They show up a lot in physical processes. You can think of the eye of a hurricane as a fixed point — while everything is being blown all over the place everywhere else, it's completely calm in the hurricane's eye. If you saw the movie *A Beautiful Mind*, the story of the triumphs and challenges of the mathematical economist John Forbes Nash, you saw the story of a man whose work on fixed points won him a Nobel Prize.

When Kaprekar wrote his paper, there were only a handful of computers in the world, and they were very expensive. So Kaprekar had to do all his work the old-fashioned way — by hand. But nowadays computers are very cheap. This book is being written on a computer which cost considerably less than a thousand dollars — and if you program it correctly, it will find the Kaprekar number in less than a second.

Here's a question — is there an octal Kaprekar's number? I confess I don't know, but I suspect there is. However, it's a lot harder, from a mathematical standpoint, to show that the Kaprekar routine always ends up in a fixed point after eight or fewer iterations of the process than it was to understand why the process that ended up with 1,089 did so.

Remember that existence proof we saw earlier concerning Taxicab numbers? We can give a somewhat similar proof to show that there are a whole lot of processes which behave similarly — after a maximum number of steps (eight in the Kaprekar routine), you eventually reach a fixed point. However, the problem is finding an algorithm — a set of computational rules — for finding that process.

A computer enables you to test out any such rule, provided you have rudimentary programming skills in a computer language such as Basic, which is the one I use (and you can learn the basics of Basic in a few hours). So here's your shot at fame and fortune — well, fame anyway. Just devise a computational routine which, when you input a four-digit number gives you another four-digit number. Kaprekar arranged the digits in ascending and descending order and subtracted them. How about simply starting with a four-digit number, moving the first digit to the last position, and subtracting the smaller of the two from the larger? Let's see what happens if we start with 1,729.

1,729	7,291
	1,729
	5,562
5,562	5,625
	5,562
	63
63	630
	63
	567
567	5,670
	567
	5,103
5,103	5,103
	1,035
	4,068
4,068	4,068
	684
	3,384

Ok, after five iterations we haven't seen anything unusual. But that doesn't mean we won't. And one thing we can be sure of — this process will eventually get stuck in some sort of loop, because there are only 10,000 four-digit numbers (from 0000 to 9,999), and sooner or later we'll see one of these numbers a second time. And when we do, the process will obviously have to repeat.

The question is — will mathematicians (or anyone else) find the results intriguing? Well, mathematicians find a lot of results intriguing that others might not, so why not take a few days, learn Basic, and see what you can come up with?

It's pretty unlikely that you will turn out to be the next Ramanujan. Geniuses like Ramanujan are rare. But with a little bit of luck, you might be the next Kaprekar.

Beating the Averages in Investing

You might ask how this topic snuck into a book on the seductiveness in math. Well, if there's anything that's more seductive than math, it's money (and yes, I know there are a couple of other candidates out there). And I absolutely had to put this in the book, and it fits nicely here.

I grew up with averages — baseball batting averages. There wasn't a scroll on the screen to give you all the numbers they do these days — actually, there wasn't even a screen. We listened to baseball on the radio and read about it in the paper the next day. And there are a lot of other averages in baseball — fielding percentages, slugging percentages, Earned Run Averages, that taught us a lot about the arithmetic of averages.

This is very useful if you get into any STEM subject because a whole lot of science is built on the math that surrounds averages. If there were a poll taken among mathematicians, scientists, and engineers as to the most useful math topic, I'm guessing averages would be up near the top.

The concept I am going to discuss is called 'dollar cost averaging'. When I stuck that term into Google, it got over 9,000,000 hits in 0.70 seconds, so it's not exactly a secret. But it's easy to illustrate.

Let's say you and your friend have both decided to invest in Cryptocorn, a stock that combines the mystery of cryptocurrency with the magic of unicorns. Cryptocorn is at $10 a share now. Your friend decides to buy 10 shares every week. Here's a chart of the share price on the Monday of each week; your friend had entered an order to buy 10 shares at the open.

Time	Now	Week 1	Week 2	Week 3	Week 4
Share Price	10	12	10	8	10
Friend Pays	100	120	100	80	100
# Friend's Shares	10	10	10	10	10

You, however, had entered a different order. Yours was to buy $100 worth of stock. Here's what the table looks like for you.

Time	Now	Week 1	Week 2	Week 3	Week 4
Share Price	10	12	10	8	10
You Pay	100	100	100	100	100
# Your Shares	10.00	8.33	10.00	12.50	10.00

You both paid 500 — but your friend's stock is worth 500, while yours is worth 508.33.

OK, it doesn't look like all that much, but you made a profit where your friend broke even. One way to measure the effectiveness of the two investing strategies is to look at the average share price of the shares you own. Your friend owns 50 shares and paid 500 for them — an average of 10 a share. You own 50.83 shares and paid 500 for them, an average share price of 9.84,

You both bought Cryptocorn because you felt it had potential. And the magic of unicorns combined with the mystery of cryptocurrency mesmerized the market over the next few weeks, as the following table will illustrate.

Time	Now	Week 1	Week 2	Week 3	Week 4
Share Price	10	11	12	13	12
Friend Pays	100	110	120	130	120
# Friend's Shares	10	10	10	10	10
You Pay	100	100	100	100	100
# Your Shares	10	9.09	8.33	7.69	8.33

And here's the bottom line.

	Friend	You
Amt Invested	580	500
# Shares Owned	50.00	43.45
Cost/Share	11.60	11.51

You've done better — at least in the sense that you bought your shares at a lower price than your friend.

But suddenly the market turns. Maybe investors have decided that cryptocurrency is too mysterious, or unicorns are insufficiently magical. Who knows? Investors are fickle. But here's how you and your friend did.

Time	Now	Week 1	Week 2	Week 3	Week 4
Share Price	10	9	8	7	8
Friend Pays	100	90	80	70	80
# Friend's Shares	10	10	10	10	10
You Pay	100	100	100	100	100
# Your Shares	10	11.11	12.50	14.29	12.50

And the all-important summary.

	Friend	You
Amt Invested	420	500
# Shares Owned	50.00	60.40
Cost/Share	8.40	8.28

It doesn't matter whether you are in an up market or a down market — you will spend less per share if you invest the same amount of money each week than if you just bought at the day's price.

And that's dollar cost averaging. It won't make you rich — but I've simply done what MANY seasoned investors do. Pick an average which goes up over time, such as the Dow-Jones or S&P. This is a long-haul strategy, so think of it as something you might like to start when you are young (if you have that option), it will enable you to retire sooner.

Of course, as you start making more money, you may want to increase the monthly amount. You might decide to invest 200 weekly, or 100 twice a week. It shouldn't make much difference, but as a matter of taste I prefer to invest 100 twice a week, as it makes my investments mirror more closely the movement of whatever I am investing in.

And you'd be amazed at how effective this can be when you are investing in something that doesn't move much over the long haul, but is very volatile in the short term. Let's look at the basic tables one more time for a really volatile investment.

Time	Now	Week 1	Week 2	Week 3	Week 4
Share Price	10	20	10	5	10
Friend Pays	100	200	100	50	100
# Friend's Shares	10	10	10	10	10
You Pay	100	100	100	100	100
# Your Shares	10.00	5.00	10.00	20.00	10.00

As I'm sure you can guess, you're really doing a whole lot better here.

	Friend	You
Amt Invested	550	500
# Shares Owned	50.00	55.00
Cost/Share	11.00	9.09

The price is exactly where it was 5 weeks ago, you paid $500 for your 55 shares, but your friend paid $550 for 50 shares — and while he has lost money you have made a profit of 10% on your investment. And even if the price of a share suddenly drops to 9.50, your friend's portfolio is worth 475, whereas yours is worth 522.50. Even though the price has dropped from the time you started doing this, you'll show a profit as long as the price stays above 9.10.

If nothing else, this should justify the investment you made buying this book.

Bibliography

[1] A. Benjamin, *The Magic of Math.* New York NY: Basic Books, 2016.

[2] MacTutor. Online at https://mathshistory.st-andrews.ac.uk/Biographies/Hardy/.

[3] MacTutor. Online at https://mathshistory.st-andrews.ac.uk/Biographies/Ramanujan/.

[4] MacTutor. Online at https://mathshistory.st-andrews.ac.uk/Biographies/Euler/.

[5] **Proof:**

Let N be a large positive integer to be specified later. Consider all systems of positive integers $(a_1, a_2, \ldots, a_{n+1})$ such that $N >= a_1 >= a_2 >= \cdots >= a_{n+1}$. There are at least

$$s = N^{n+1}/(n+1)!$$

such systems, as there are N^{n+1} $(n+1)$-tuplets of positive integers not exceeding N, and any nonincreasing sequence corresponds to at most $(n+1)!$ such tuplets. For each of the systems, consider the sum

$$a_1^{\,n} + a_2^{\,n} + \cdots + a_{n+1}^{\,n}$$

being a number not exceeding

$$S = (n+1) * N^n$$

For large N (namely $N > (n+1) * (n+1)!$) we have

$$s > S$$

and there are more considered systems than possible sums associated with them. By pigeon-hole principle there exist two different systems having the same sum associated with them, giving

$$\boldsymbol{a_1^n + a_2^n + \cdots + a_{n+1}^n = b_1^n + b_2^n + \cdots + b_{n+1}^n}$$

with

$$\boldsymbol{N >= a_1 > = a_2 >= \cdots >= a_{n+1}} \quad \text{and} \quad \boldsymbol{N >= b_1 >= b_2 >= \cdots >= b_{n+1}}.$$

Although systems $\boldsymbol{(a_1, a_2, \ldots, a_{n+1})}$ and $\boldsymbol{(b_1, b_2, \ldots, b_{n+1})}$ are different, they may have some common elements. QED.

[6] D. Kaprekar, *An Interesting Property of the Number 6174*. New York, NY: Scripta Mathematica, 1955.

Chapter 3

Seduced by Patterns

According to the dictionary, a pattern is a repeated or regular arrangement — and visual and symbolic patterns have fascinated human beings since before we were even able to articulate the idea. The Sun rises in the East and sets in the West. In the Northern hemisphere, cold winter is followed by warm spring, hot summer, cool fall, and cold winter in endless succession.

And of course, there are those "fill in the next number" patterns we saw in elementary school tests. Most of us handled the next number after 1, 2, 3, 4, 5,... without too much difficulty. Likewise for 2, 5, 8, 11, 14,..... Maybe 1, −2, 4, −8, 16 ... gave us a little trouble — but probably not if you're reading this book.

Recursive and Explicit (Formulaic) Patterns

Mathematicians have defined a sequence as a collection of things placed in one-to-one correspondence with the positive integers. If n is a positive number, we write the nth thing as a_n. It's possible for "things" to be anything whatsoever — but let's assume for the moment that we're letting these things be numbers. Take a look at that last sequence, the first five terms of which are 1, −2, 4, −8, 16.

The first thing we might notice is that each term is −2 times the preceding term. Using subscripts, we write this as $a_{n+1} = -2a_n$. That doesn't completely specify the sequence at which we are looking, as the same thing could be said for the sequence that starts out −3, 6, −18, 36, −72. But if we specify that the first term $a_1 = 1$, the statement

$$a_{n+1} = -2a_n \quad \text{and} \quad a_1 = 1$$

gives us all the information we need to completely describe the sequence. This is known as a recursively defined sequence — we know where to start, and we know how to find the next term.

However, if that's all we have, and we're asked to find the 19th term of that sequence, it's going to take us a little while. To be precise, we'd have to work our way through the first 18 terms to be able to calculate a_{19}.

So what we'd prefer is an explicit formula for a_n, one which enables us to compute a_n directly without the aggravation of having to compute a lot of other values before we get to the one we want. And in this instance, it's pretty easy to see what we want. We start with a_1, and then every term thereafter simply multiplies the result by −2, so by the time we get to a_n we'll have multiplied by −2 a total of $n-1$ times. As a result, we conclude that

$$a_n = a_1(-2)^{n-1}$$

In our case, when $a_1 = 1$, the formula reduces to $a_n = (-2)^{n-1}$.

There's always a formula associated with every recursive definition simply by defining a function f on the positive integers such that $f(n) = a_n$ — but that formula may not be one that's usable in terms of making our computational lives easier. A "nice" formula would be one that strings together a finite number of variables and recognizable mathematical operators — but that may not always be possible. Now we're starting to venture into the realm of something called computability.

We've already examined a very important sequence back in the first chapter — the sequence of prime numbers. At this moment, it is not known whether the sequence of primes is computable. There are a number of formulas which generate a sequence of nothing but primes, but it is not known whether there is a formula which will generate every prime — or even whether such a formula exists.

There are a number of problems such as this one, in which the goal would be to find an explicit solution to a problem. Failing that, it would be nice to know whether or not such an explicit solution exists — it would at least tell us that there's a pot of gold at the end of the rainbow.

Fibonacci Numbers

Back about 800 years ago, there lived perhaps the greatest mathematician — at least in Europe — of the Middle Ages. He was actually named Bonacci, and was the son of someone named Bonacci, but an historian of the era gave him the nickname Fibonacci (short for *filius Bonaccci*, son of Bonacci), and he's been stuck with it ever since.

Anyway, he propounded a relatively easy to solve puzzle that has probably kept more mathematicians busy than any other single result in mathematics — if only because they've had 800-plus years to think about it. Fibonacci postulated a species of idealized rabbits that were born, took a month to reach maturity, and produced another pair of idealized rabbits on the second day of the month until the end of time. The question Fibonacci asked was — if one put one pair of idealized rabbits in a field subject to this hypothesis, how many pairs of idealized rabbits would there be on the first day of the nth month?

Nowadays this problem is stated mathematically as follows. The nth Fibonacci number F_n is defined recursively by

$$F_0 = 0 \quad F_1 = 1 \quad F_n = F_{n-1} + F_{n+2} \quad \text{for } n \geq 2$$

Fibonacci might have thought of the definition of F_0 as 0 and F_1 as 1 as the field being initially empty, and 1 month later the pair of idealized rabbits is placed in it. But he certainly would have recognized where the recursive relation comes from. The number of rabbits present certainly includes all the ones who were there on the first of the previous month — that's F_{n-1}. But on the second of the previous month, all the rabbits who were there on the first of the month before that have another pair of rabbits — and that's F_{n-2}. Adding the mature rabbits to the not-yet-mature rabbits gives us the recurrence formula.

Here are the first few Fibonacci numbers.

$$0, 1, 1, 2, 3, 5, 8, 13, 21, 34, 55, 89, 144, 233, 377, \ldots.$$

This is a fascinating sequence of numbers, and amazingly enough, it appears with some frequency in Nature. The interested reader can easily find the details, but Fibonacci numbers occur in a variety of fruits, flowers, vegetables and trees as the number of florets in spirals, the number of petals on flowers, or the number of leaves on a tree.

Certainly of interest to botanists, but of more interest to mathematicians was what happened when they looked at the ratios of successive terms. Here are decimal approximations to four decimal places.

$$2/1 = 2.0000$$
$$3/2 = 1.5000$$
$$5/3 = 1.6667$$
$$8/5 = 1.6000$$
$$13/8 = 1.6250$$
$$21/13 = 1.6154$$

It sure looks like those ratios are headed somewhere. In order to compute that "somewhere", I'm going to ask you to make a computational approximation which mathematicians did routinely until the early 19th century, when they found out this approximation could get them into trouble. The approximation is this: there is a number X such that after we go far enough out in the sequence, the ratio of successive terms is always X. This might happen — or it might not — but for the time being let's go with that assumption.

Then if we go far enough out in the sequence, say F_N (N is clearly a VERY large integer), then $F_{N+1}/F_N = X = F_{N+2}/F_{N+1}$. From the first equality, we get $F_{N+1} = XF_N$, and from the second equality we get $F_{N+2} = XF_{N+1} = X^2F_N$. Substituting these expressions into the recurrence relation, we get

$$F_{N+2} = F_{N+1} + F_N$$
$$X^2 F_N = X F_N + F_N$$

Dividing both sides by F_N, we see that $X^2 = X + 1$. Writing this as $X^2 - X - 1 = 0$ and applying the quadratic formula, we see that

$$X = \frac{1 \pm \sqrt{5}}{2}$$

Since the ratio always has to be positive, we have

$$X = \frac{1 + \sqrt{5}}{2}$$

This is another number that crops up all over the place in mathematics. It's known as the Golden Ratio, and is abbreviated as Φ (the Greek letter capital phi, or Phi).

Excel 101

I think mathematics today is more enjoyable than it was when I was growing up, and that's largely because of computers. I've talked previously about how being able to write simple programs in Basic enriches one's ability to appreciate certain aspects of mathematics.

But you can do a lot with Excel, or any similar spreadsheet. In the material that comes next, I'm assuming a familiarity with some elementary aspects of Excel, such as writing formulas, filling cells ("fill" is what Excel calls placing a repeated formula in cells) and making charts. There's plenty of material online, as these are very basic things to do.

Take the Fibonacci numbers, and open your Excel spreadsheet. Enter 0 into cell A1 (this is the desired starting value for F_0) and 1 into cell A2 (this is the desired starting value for F_1). In cell A3, enter

$$= \text{A1} + \text{A2}$$

And in cell B3, enter

$$= \text{A3}/\text{A2}$$

Now fill down cells A3 and B3 to row 25. Cells A1 through A25 will contain the first 25 Fibonacci numbers, and cell Bk will contain the quotient F_k/F_{k-1}; notice how rapidly it stabilizes at the first digits of Φ.

Change the numbers in cells A1 and A2 to get different starting populations and notice that even though cells A3 through A25 differ from what they were before, the cells in the B column again stabilize at the first digits of Φ. Of course, algebra told us this would happen, but it's nice to get confirmation!

Fun with Fibonacci — Excel 102

There's actually a Fibonacci Association. It's been in existence since 1964, and there's also a *Fibonacci Quarterly* [1] journal, which publishes articles related to guess what? Some are pretty arcane — but you can do an awful lot with very little knowledge and your trusty Excel spreadsheet.

Suppose we wanted to explore the following question: What happens if we decide to generalize Fibonacci's basic idea in the following possible directions.

(1) Finite lifetime — instead of being immortal, as Fibonacci specifies, each pair of rabbits lives for only a finite length of time. One could either specify a common lifetime for each pair of rabbits, or a mortality distribution for each generation of rabbits. Or other possibilities.

(2) Reproduction interval — instead of reproducing monthly, as Fibonacci specifies, each pair of rabbits has a reproduction schedule

It is relatively easy to decide what specific modification to explore, incorporate it in the basic Fibonacci recurrence relations, and see what becomes of the quadratic equation in X to which Φ was a solution. Depending upon the complexity of the conditions you set up, finding the roots of X becomes a much more difficult problem, and you will have to rely on numerical solutions.

Or you could let Excel do it for you. You need the early cells in column A to specify the initial conditions and the recurrence relation, and make sure to enter the quotient command now in cell B3 in the cell next to the one in column A that defines the recurrence relation. Then sit back and enjoy.

Those of you who know how to insert a chart (again, a fundamental feature in Excel) can see the data translated into a chart of your preference.

The Fibonacci numbers are generally presented via a recursive formula, as we've seen, but there is actually a formula for the nth Fibonacci number. The Golden Ratio

$$\Phi = \frac{1+\sqrt{5}}{2}$$

arises as one solution to a quadratic equation; the conjugate solution is

$$\varphi = \frac{1-\sqrt{5}}{2}$$

The nth Fibonacci number is given by the formula

$$F_n = \frac{\Phi^n - \varphi^n}{\sqrt{5}}$$

This formula was presumably known to Euler (as was practically everything in mathematics prior to the 18th century) and De Moivre, but is generally attributed to the French mathematician Jacque Binet, who published it rather than just mentioned it.

OK, how would you prove it? We have two formulas for the Fibonacci numbers — the recurrence relation, plus the fact that $F_0 = 0$ and $F_1 = 1$, and the explicit formula given above. If the explicit formula gives $F_0 = 0$ and $F_1 = 1$, and if it satisfies the recurrence relation, it must be the right formula. Showing that $F_0 = 0$ and $F_1 = 1$ is trivial (mathematicians love to say this), and showing that it satisfies the recurrence relation is uninteresting algebra and is left to the reader (mathematicians love to say this as well).

The real question is — what line of reasoning led De Moivre, Euler, and Binet to come up with this formula? Mathematics is not just the formulas we see in mathematics — those are the end result. What really matters is the lines of reasoning that lead to those formulas, because those lines of reasoning can lead to other formulas — and other lines of reasoning. And that's how mathematics evolves.

More about the Golden Ratio

Even before Fibonacci — about 1,500 years before — the Greeks had come up with the idea of the Golden Ratio in connection with geometry problems. They considered a line segment of length $a + b$, divided as follows.

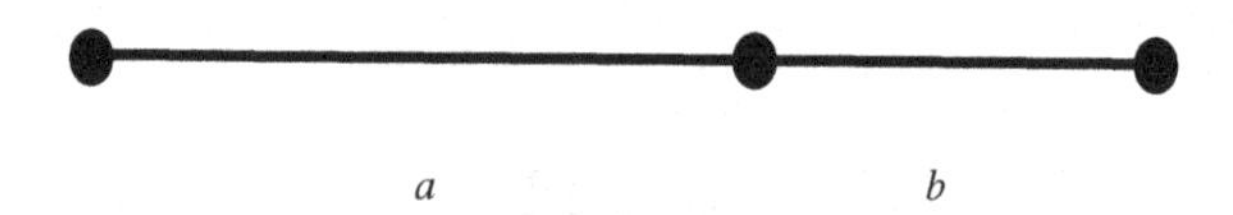

a b

This line was said to be divided into the Golden Ratio if the ratio of the length of the entire line to the larger of the two segments (a, in this case) was equal to the ratio of the larger segment to the smaller. This resulted in the equation

$$\frac{a+b}{a}=\frac{a}{b}$$

We split the left side into two fractions, and abbreviate a/b, the Golden Ratio, by Φ. This gives

$$\frac{a+b}{a}=\frac{a}{a}+\frac{b}{a}=1+\frac{b}{a}=1+\frac{1}{\Phi}$$

But since the right side of the original equation is $a/b=\Phi$, we see that

$$1+\frac{1}{\Phi}=\Phi$$

Multiplying both sides by Φ and putting all the terms on one side yields the quadratic equation which defines Φ.

$$\Phi^2-\Phi-1=0$$

The Golden Rectangle, one whose sides are in the relationship Φ:1, was felt by the Greeks to be the most beautiful rectangle — I have no idea how they came to this conclusion. However, the Parthenon in Athens is built in this ratio, but no one knows whether by design or by accident.

Johannes Kepler is generally known for his contributions to astronomy, rather than mathematics. His Three Laws of Planetary Motion were the result of more than two decades spent trying to understand the mathematics behind the orbits of the planets. But Kepler was not just a data-cruncher. Noticing that the Golden Ratio could be stated as

$$\Phi^2=\Phi+1=\left(\sqrt{\Phi}\right)^2+1^2$$

he constructed a right triangle whose sides were $\sqrt{\Phi}$ and 1 and whose hypotenuse was Φ. You might note that the three sides of the triangle in increasing order form a geometric sequence — we can be absolutely certain that Kepler, a man genuinely seduced by math, did.

In fact, it's worth discussing how the mathematical seduction Kepler experienced played a vital role in our scientific development. In Kepler's Era, only five planets other than the Earth were known: Mercury, Venus, Mars, Jupiter and Saturn. The number 5 appears prominently in solid geometry, as there are five regular solids. A regular solid is a polyhedron all of whose edges are of equal length, and such that all of the angles made by two of those edges are equal. A cube is the most obvious example of a regular solid — as we know, it has six sides, all of which are squares. It was known to the Greeks that there were only five regular solids — five regular solids, five planets, there must be some connection. Or so Kepler thought, and he burned up two decades trying to make this work out. No joy. Finally, he tried to find a model which would fit the data, and did so, using ellipses for planetary orbits instead of models suggested by a correspondence with regular solids. The Three Laws he devised could be derived from Newton's Theory of Universal Gravitation — and science has never looked back.

Kepler lived during the 15th and 16th centuries, an era in which mathematics was fascinated by continued fractions — a fascination which I confess I do not understand. However, if we notice that

$$\Phi = 1 + \frac{1}{\Phi}$$

And substitute the right-hand expression for the value of Φ in the denominator of the fraction on the right-hand side, and continue doing this, we get

$$\Phi = \frac{1}{1+\frac{1}{1+\frac{1}{1+\ldots}}}$$

as a continued fraction representation of Φ.

The Quicksand of Mathematics

Another fascinating number pattern was discovered in the 20th century by the German mathematician Lothar Collatz, Start with any positive integer greater than 1. If it is even, divide it by 2. If it is odd, multiply by 3 and add 1. If you start with 17, the pattern this rule generates is 17, 52, 26, 13, 40, 20, 10, 5, 16, 8, 4, 2, 1. The Collatz Conjecture — still unsolved as of this writing — is that no matter what positive integer you start with, you'll always eventually reach 1. I'm going to refer to the process of constructing a sequence such as above, where we started with 17, as the Collatz process.

Thanks to computers, mathematicians have been able to show that if you start with a number less than 1 quintillion — that's 1,000,000,000,000,000,000-, you do eventually reach 1. But there are a lot more numbers bigger than 1 quintillion — infinitely many. So this isn't doesn't even scratch the surface — or does it? More on that later.

The Collatz Conjecture looks like it was something discovered by a boy who liked playing around with numbers and had just learned about something called piecewise-defined functions (a function which is defined differently for at least two disjoint subsets of the domain). But beware — the Collatz Conjecture has a reputation as the quicksand of mathematics. Because it is so simple to state, anybody can start to work on it. And once someone starts work on it, it's almost impossible to stop — you keep sliding deeper and deeper.

One of the greatest mathematicians of the 20th century, Paul Erdos, said of the Collatz Conjecture that "mathematics is not yet ready for such problems." That could have been both a statement about the total intractability of the problem, but also a veiled warning that working on it could prove a fruitless endeavor.

But that didn't dissuade people from trying. So what follows is a progress report — because progress has been made. In the 1970s, mathematicians managed to show that almost all sequences (there's a precise definition for "almost all" — because there's a precise definition for everything in mathematics — but we needn't go into it here) eventually reach a number that's smaller than the starting number.

It would be tempting to say that we've done most of the work already with this result. Start with any number that belongs to the "almost all" mentioned in the previous paragraph, call it N_1. The Collatz process eventually reaches a number, which we'll call N_2, which is less than N_1. Start with N_2, and apply the Collatz process. It eventually reaches a number N_3, which is less than N_2. Since there are only $N_1 - 1$ numbers smaller than N_1, sooner or later we'll reach 1. So it appears that the Collatz process reaches 1 for almost all numbers.

I don't think there's a mathematician alive who at one time or another has come up with an argument for a desired result that falls apart on closer scrutiny, and closer scrutiny of the previous paragraph will cause its argument to fall apart — and we can see exactly where it falls apart. Even though N_2 is less than N_1, we have no assurance that N_2 belongs to the set of "almost all" integers which eventually reach smaller numbers than the one at which we started. So we can't necessarily find N_3, because N_2 might just be a number for which the Collatz process never terminates.

Remember the Twin Primes Conjecture, where no progress was made for quite some time, and then Zhang managed to show that there were infinitely many pairs of primes less than 70 million apart? The gap of 70 million has been greatly narrowed since then, and one of the mathematicians who did yeoman work narrowing the gap was Terence Tao. Somehow Tao got interested in the Collatz Conjecture — and managed to avoid getting caught in the quicksand by simply saying something on the order of, "I'll give myself a few days to look at it, and if I'm not getting anywhere, I'll move on."

Over the past decade, Tao spent the allotted few days — several times, but never made any notable progress. And this is just the way mathematics is — as every living mathematician can attest. But then, Tao found a way to strengthen the result obtained in the '70s. Tao's result is that 99% of numbers sufficiently large initiate Collatz processes that get very close to 1. It's on the order of "99% of all numbers larger than 10^{100} initiate Collatz processes that eventually get smaller than 10^{10}." But I used 10^{100} and 10^{10} as proxies for numbers that Tao showed existed, but whose value has not been computed — and indeed may never be computed [2].

Mathematics is very much like fishing — the biggest fish are generally the hardest to land. There's a reason that the Twin Primes Conjecture and the Collatz Conjecture are so notorious — they've been around for a while, and progress is sporadic.

But every so often, someone lands a Really Big Fish. It was headline news in the 1990s when Andrew Wiles managed to prove Fermat's Last Theorem — more than three centuries after it had been posed. Possibly someone will come up with an ingenious idea that finally resolves the Twin Primes Conjecture or the Collatz Conjecture.

If there were a gambling casino on which one could bet on the truth or falsehood of mathematical conjectures, I have no idea how I'd bet on the Twin Prime conjecture. But I would shove my chips all in on the Collatz Conjecture being true. At its heart, it's just a conjecture about the numbers 2 and 3, and it sure seems to me that if every number smaller than 1,000,000,000,000,000,000 initiates a Collatz process than ends in 1, they all do.

Seduced by Hidden Patterns

I don't think it's possible to introduce children to patterns too early, and as Bismarck once said, you can do anything with children if only you play with them [3]. So let's look at a couple of hidden pattern puzzles that I've used very successfully when I go to an elementary school.

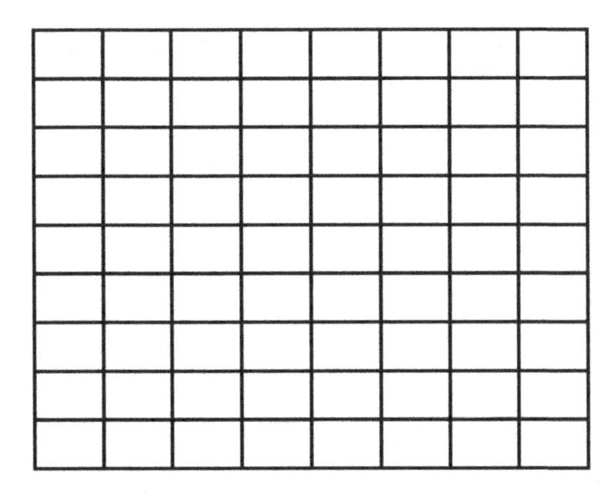

This is an 8×8 grid. I give each child one of these, along with a large supply of 2×1 tiles that look like this

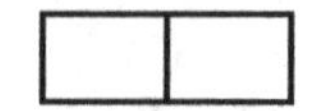

I then ask them to cover the 8×8 grid with the tiles so that the grid is completely covered and no 2 tiles overlap each other or extend outside the grid. They usually have no difficulty doing this. I then remove one of the corner squares so it looks like this.

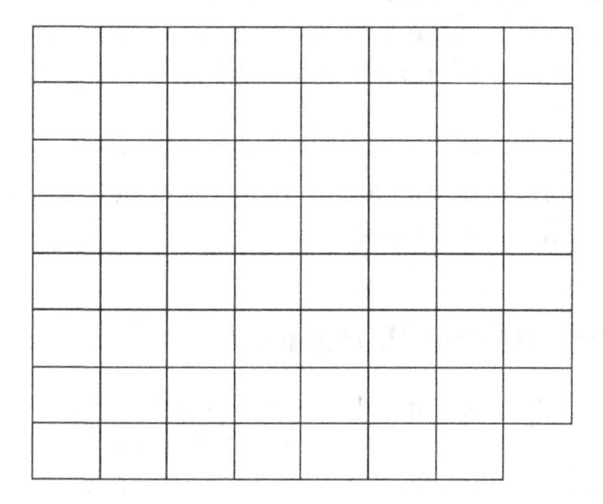

Some of them see fairly quickly that you can't do this, because each tile covers 2 squares, and there are an odd number of squares in the above diagram. After everyone understands this, we move on to the next problem, in which I remove the square in the upper left-hand corner. The grid now looks like this.

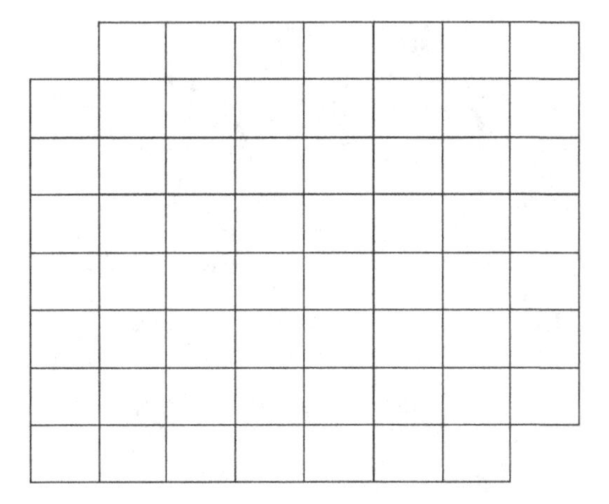

Now the experimentation begins in earnest. After a few minutes, I call a halt to the proceedings and tell them that it might be an easier problem if we did it with a regular checkerboard.

Here's what the checkerboard looks like originally.

If we remove the upper left and lower right-hand corners, it looks like this.

And we use tiles that look like this.

Sooner or later, someone sees that each tile covers one black and one gray square, but the checkerboard with the corners removed has 32 gray squares and 30 black squares, so you can't do it.

Last Marble in the Jar

Here's another one — I bring the equipment to the class, but I'm going to ask you to visualize it. A jar contains 21 white marbles and 21 black marbles. You reach in and select a pair of marbles — either at random or by choosing them. The rules are:

(1) If both marbles are the same color, remove them from the jar.

(2) If the marbles are of different colors, remove the black marble and put the white marble back in the jar.

Question — what color is the last marble remaining in the jar?

It's not immediately clear that there will be a last solitary marble in the jar, but you can see that each move reduces the number of marbles in the jar by one or two marbles, so sooner or later you'll be down to either two or three marbles. However, the hidden pattern here is to note that either we remove two white marbles or no white marbles. That means that when the number of white marbles is reduced, it's always reduced by 2. So there can never be zero white marbles in the jar, since we started with 21, and so the last marble must be white.

It's not easy to spot a hidden pattern. Duh, it's hidden — as in "hard to find". You are to be congratulated if you see ANY hidden pattern. The important point is to recognize what the pattern gives you in terms of the problem.

A Different Last Marble Problem

Another problem involving marbles features a jar with six marbles of assorted and completely irrelevant colors. You and an opponent are playing a game in which the object is to remove the last marble. You each play alternately, and must remove either one or two marbles. Would you rather go first or second, and why?

There's a temptation to analyze this by starting with 6 marbles, and investigating the possible games if you go first or second, and take one or two marbles. That's four possible decision trees — where you go first and take two marbles, etc.,- and each one of the four trees has a number of possible games to consider. This method — perfectly reasonable, BTW — is going to keep you here awhile.

A much better approach is to analyze the game at its conclusion. If it's your turn to play and there are only 1 or 2 marbles remaining, you're obviously a winner — you just remove all of them. But what if there are 3 marbles in the jar?

If it's your turn to play, you can see quickly that you're going to lose, as no matter what you do, your opponent will be confronted with a jar that has only 1 or 2 marbles. So obviously you want it to be your opponent's turn to play when the jar contains 3 marbles.

A moment's thought will convince you that if your opponent makes the first move when the jar contains the original 6 marbles, he or she is doomed — if he or she removes x marbles, you simply remove $3 - x$. It will then be your opponent's move with 3 marbles remaining, making you a winner.

But there's a pattern here that enable us to go beyond the 6 marble jar. If the jar contains a multiple of 3 marbles, you let your opponent go first, and simply make sure that the total number of marbles removed by you and your opponent together is 3. That way, the total number of marbles in the jar is always reduced by 3, and sooner or later your opponent finds that the jar only contains 3 marbles.

In fact, the same reasoning shows that if the jar does NOT contain a multiple of 3 marbles, you want to go first — that is, as long as you know the number of marbles in the jar. You simply remove either 1 or 2 marbles to reduce the total number of marbles to a multiple of 3. And that brings us to the mathematician — who is again applying for a job.

The interviewer shows the mathematician to a room in which a fire is burning, and tells him to put out the fire. There is a table with a bucket of water on it. The mathematician grabs the bucket and dumps water on the fire until it is out.

"Do I get the job?," the mathematician asks.

"That was only the preliminary test," the interviewer responds. "You passed — but now for the advanced test." He takes the mathematician to a room in which a fire is burning and — surprise! — tells him to put out the fire. There is a table with a bucket of water under it. The mathematician grabs the bucket and — puts it on the table.

The interviewer is stunned. "Why did you do that?" he asks.

"To reduce the problem to one I've already solved," the mathematician replies.

Reducing a problem to one that has already been solved is one of the major reasons that math continues to grow. But another reason for its growth is that an important part of mathematics is the search for — and the application of — patterns, because that enables us to go beyond simple problems to more complex ones.

We started with a jar containing 6 marbles, and the game consisted of each player alternately removing one or two marbles until the winner removes the last marble. Suppose now we have a jar containing a known number of marbles N, and each player must remove at least one marble, but may remove as many as p marbles. Take a moment to see if you can determine whether you want to go first or second, and how you want to play the game.

Seeing the pattern for playing the game in the case when you must remove one or two marbles makes it clear how to go about this game. If it is your opponent's turn to play when there are $p + 1$ marbles in the jar, you are a winner. And you can arrange for this to happen in one of two cases.

Case I — If there are a multiple of $p + 1$ marbles in the jar, make your opponent go first. If he removes x marbles, you remove $p + 1 - x$. Sooner or later, it will be your opponent's turn to play with $p + 1$ marbles in the jar.

Case II — If there aren't a multiple of $p + 1$ marbles in the jar, you should go first, and remove enough marbles so that there are now a multiple of $p + 1$ marble remaining in the jar. Mathematicians use the notation

$$x \bmod y$$

to denote the remainder of x on division by y (example: $18 \bmod 7 = 4$). Using this notation, if $N \bmod (p + 1) > 0$, go first and remove $N \bmod (p + 1)$ marbles. Now the jar has a multiple of $p + 1$ marbles, and we've reduced Case II to Case I — and you're a winner.

The physicist Eugene Wigner once wrote an article [4] about why mathematics was so effective a tool for science. One of the reasons is that Nature creates patterns — if everything in Nature were patternless and disorganized, there would be no Nature to study, and no human beings, either. But Nature creates patterns, and it is for us to understand those patterns and use them to improve our lives — and a large part of mathematics is the search for patterns.

Bibliography

[1] *Fibonacci Quarterly*. Online at https://www.fq.math.ca.

[2] *Quanta Magazine*. Online at https://www.quantamagazine.org/mathematician-proves-huge-result-on-dangerous-problem-20191211/.

[3] *Inspiringquotes.us*. Online at https://www.inspiringquotes.us/author/4774-otto-von-bismarck.

[4] E. Wigner, *The Unreasonable Effectiveness of Mathematics in the Natural Sciences*, Communications on Pure and Applied Mathematics. Hoboken, NJ: John Wiley & Sons, 1960.

Chapter 4

Seduced by Analytic Geometry

I really wanted to write a chapter entitled "Seduced by Geometry." I really did, because probably more "outsiders" — people who aren't seriously connected to math in their careers — have been seduced by geometry than by any other branch of mathematics. Geometry has seduced Presidents, kings, and emperors, and I don't think any other branch of mathematics can make that statement.

But it's really hard to write about something you haven't experienced — and as I mentioned in the Introduction, I never really came to love geometry. I came to appreciate its fundamental role in mathematics — just as one can appreciate the fact that Bach was a great composer even if most of his music leaves you cold. And that's a really good analogy in my case, because most of Bach leaves me cold.

But not all of Bach. There's some of his stuff I find quite enjoyable — and the same can be said of geometry. So let me show you what for me are a couple of the high points of geometry.

The Unexpected Parallelogram — Varignon's Theorem

A simple, yet surprising result, can be found by drawing a convex quadrilateral. "Convex" means that if you join any two points inside the shape by a line segment, the entire segment stays completely inside the shape. Triangles, squares, and circles are convex — the inside of a figure-eight isn't, because a line between two points in the two separate circles that make up the figure-eight strays outside the figure-eight. Take the midpoints of each side of the quadrilateral, and connect each midpoint by a line segment to the midpoint of each of the two adjacent sides, as shown in the following figure. Amazingly enough, the lines connecting the midpoints form a parallelogram — no matter what convex quadrilateral you start with.

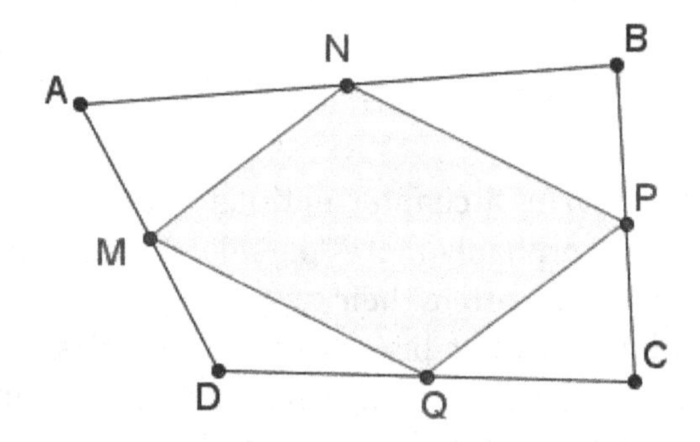

And now we come to the crux of what troubled me in geometry (as it troubles a lot of geometry students) — proof. How do we prove that result?

I'm not sure that I could prove it, but you can find anything online these days, so here's a proof I found online.

Referring to the diagram above, ABD and AMN are similar by the side-angle-side criterion, so the angle ADB and AMN are equal, making *MN* parallel to DB. In the same way, PQ is parallel to BD, so MN and PQ are parallel to each other; the same holds for MQ and NP.

First, props to you if you found that proof. I'm not sure I could have if left to my own devices. Second, even if I had found that particular proof, my geometry teacher would have probably given me a B. In the geometry course I took, side-angle-side was a theorem about congruent triangles, rather than similar ones. I'm sure the relevant theorem about similar triangles was in the book, but I don't think it was called that. Additionally, phrases like "the same holds" never saw the light of day in my geometry course. But this proof illustrates one of the ways I think that the teaching of geometry has improved — it's not necessary to dot every i and cross every t the way it was back in the day.

Morley's Theorem

There are any number of "unexpected triangle" theorems in geometry such as Morley's Theorem, which states that the angle trisectors of any triangle meet in an equilateral triangle. This was only unearthed in 1899; here's a picture.

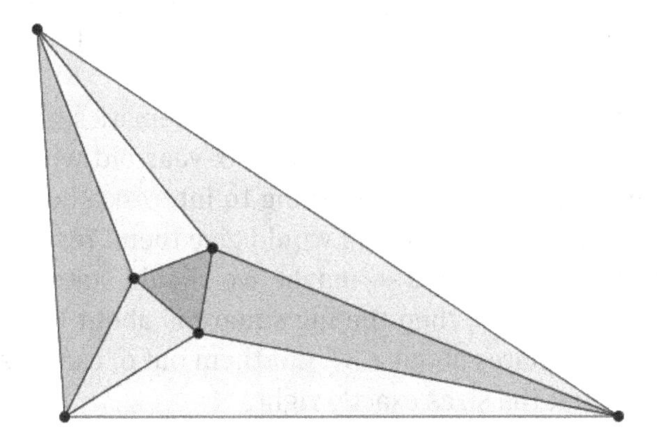

I'm told a geometric proof is brutal. Frankly, although others say this is very seductive, for me it's in the "that's nice" category — at best.

But because I don't want to leave geometry without mentioning the one problem that REALLY piques my curiosity, let me show you this. It's simple to state as it involves nothing but squares.

Covering Up

Suppose you have a square that is 201 inches by 201 inches. Its area is 40,401 square inches. Suppose you also have six squares that each measure 100 inches by 100 inches. Each square has an area of 10,000 square inches, so the total area of all six squares is 60,000 — seemingly more than enough to cover the 201×201 square. And it would be — if you were allowed to cut up the 100×100 squares. But suppose you're not — you have to try to arrange the six 100×100 squares to cover the larger square. Can you do it?

As of this writing, no one knows. If you use seven 100×100 squares you can do it — but no one knows how to do it with six 100×100 squares — or is able to show that it's impossible. The brilliant mathematician Maryam Mirzhakani, the only woman to win the Fields Medal — mathematics' highest award — who sadly passed away at age 40 in 2017, would study geometry by making paper and cardboard cutouts of problems like this and moving the shapes around. One of the most seductive things about mathematics is that you can conduct explorations with the simplest of tools — pencil, paper, and cardboard cutouts.

This is the only unsolved problem in mathematics that I know of which could conceivably be solved by a six-year-old with no mathematical training. In fact, if I were trying to interest a boy or a girl in mathematics, this is the problem I would give them. Yes, it's probably a pain making those squares — maybe we should state the problem in terms of centimeters, then the big square is about 6′7″ on a side and the smaller squares about 3′3″. Cut them out of cardboard — but make sure you get the sizes exactly right.

Homage to the Greeks

The fact that I haven't really been seduced by classical geometry the way the Greeks saw it and the way it is generally taught today in geometry classes in no way diminishes my respect and admiration for the Greeks who developed this subject.

I'm genuinely amazed at what they were able to do — especially considering that they didn't even have the equivalent of scratch paper to work on, as far as I know. They only had a few scrolls — so that means that most of what one Greek knew about geometry had to be learned from another Greek who knew about geometry. They also didn't have anything like arithmetic and algebra as we know it today, and when you think of what they were able to accomplish with these handicaps, it's even more impressive.

And when you consider their accomplishments in solid geometry, it's little short of miraculous. I've taken a lot of math courses at all levels of instruction in my life, and the single most difficult course I ever took was solid geometry in high school. I'm guessing I wasn't the only one who found it difficult, as this course is now basically defunct — it's not taught anywhere as part of the mainstream curriculum. And there's good reason for it — calculus does a vastly better job, and while plane geometry helps prepare for calculus, solid geometry doesn't.

But the Greeks got many of the important results without calculus. I've never read a classical geometry proof of the formula for the volume of a sphere — to me one of the first seductive things I found in calculus was a 30-second derivation of that formula.

But never, ever, would I disparage the Greek mathematicians. Those guys were GOOD — and we should never forget it.

Descartes and the Birth of Analytic Geometry

Rene Descartes was a French philosopher, scientist, and mathematician. Although he is undoubtedly best known for the phrase, "I think, therefore I am," my personal opinion is that the total impact on human progress of all of his philosophical works doesn't even come close to the value of the couple of pages of mathematics he appended to one of those incredibly long tracts on philosophy.

In those two pages, Descartes outlined the essence of what we now think of analytic geometry, a truly brilliant innovation that enabled the power of algebra to be brought to bear on problems of geometry — and a whole lot else.

One of the things that makes analytic geometry seductive — at least for me — is its ability to present incredibly simple arguments in place of some of the more contorted proofs you find in geometry. Here's a simple example — prove that a tangent to a circle is perpendicular to the radius at the point of contact. All the geometric proofs I found were indirect — they established the result by assuming that a negation of some sort was true, and then derived a contradiction to that negation.

Now this is an extremely valuable proof technique that appears in lots of mathematics. It's the mathematical equivalent of Sherlock Holmes' argument about the dog that DIDN'T bark in the night. But it's difficult for students to grasp at first, and so we're faced with a pedagogical question — is it better for a student to learn about indirect proofs early in the game, or learn how to prove something more directly, more easily and, yes, more seductively?

And you probably can guess on which side of that question I come down.

So here's the proof, courtesy of analytic geometry. Assume that the circle has radius 1 and is centered at the origin — this is generally called the unit circle. The radius is the line segment from the origin to the point (1,0). It is perpendicular to the line $x = 1$ at the point (1,0), and that line is tangent to the circle at that point because any point on that line OTHER than (1,0) has coordinates $(1, y)$ with $y \neq 0$. The distance of $(1, y)$ from the origin is greater than 1, so no other point besides (1,0) on the line $x = 1$ lies on the unit circle.

You might think that this proof suffers from the fact that I chose a very specific circle — the unit circle — and a very specific radius of that circle. Not at all — the circle given in the original problem (show that the tangent to a circle is perpendicular to the radius at the point of contact) is lying somewhere in space. I CHOSE the coordinate system so that the origin is at the center of the circle, the x-axis lies on top of the radius, and the length of the radius is 1 because I'm defining the unit of measurement in my coordinate plane to be the length of the radius.

In order to make my point, let me give a couple of other proofs involving circles that are considerably more difficult to construct in classical geometry. And part of the seductiveness of analytic geometry is that you don't really have to do proof construction — what you often have to do is simply algebraic verification using some of the basic ideas in analytic geometry.

How about this one — a triangle whose hypotenuse is the diameter of a circle and which has the other vertex on the circumference of that circle is a right triangle? Easy in analytic geometry, as long as you know that two lines, neither of which is parallel to one of the axes, are perpendicular if the product of their slopes is –1. This fact comes courtesy of the Pythagorean Theorem, on which analytic geometry relies quite heavily. But that's OK, the Introduction to this book has what I consider to be a seductive proof of that result.

Anyway, let's get back to the problem at hand. Let's use the unit circle; its diameter extends from the point (–1,0) to the point (1,0). Let (x, y) denote the other vertex of the triangle. The side of the triangle from (–1,0) to (x, y) lies on the line with slope $(y-0)/(x-(-1)) = y/(x+1)$, the side of the triangle from (x, y) to (1,0) lies on the line with slope $(0-y)/(1-x) = y/(x-1)$. The product of the slopes of these two lines is therefore $y^2/(x^2-1) = (1-x^2)/(x^2-1) = -1$, and we're done.

One for the road. Take any point on the unit circle not lying on a diameter, and drop a perpendicular to the diameter. The length of the line segment from (x, y) to the diameter is the geometric mean (the square root of the product) of the lengths of the two line segments from (–1, 0) to $(x, 0)$ and from $(x, 0)$ to (1, 0). The classical proof is usually accomplished by invoking theorems about the ratios of corresponding sides in similar triangles, but it requires some agility to figure out which triangle is similar to which. But the analytic geometry proof is a simple algebraic verification: by symmetry, we can assume $y > 0$. The distance from (x, y) to $(x, 0)$ is

$$\sqrt{(x-x)^2+(y-0)^2} = y$$

$$= \sqrt{(1-x^2} \text{ since}(x,y) \quad \text{lies on the unit circle}$$

$$= \sqrt{(1-x)(1+x)} \quad \text{algebra}$$

But the first expression under the radical is the length of the segment from $(x, 0)$ to (1, 0), and the second is the length of the segment from (–1, 0) to $(x, 0)$.

But It Doesn't Just Make Geometry Easier

There's a lot more to the Cartesian plane than simply providing an easier path to the proof of theorems in geometry. When connected with the idea of function — using a graph to display the curve $y = f(x)$ — it is one of the most powerful and productive ideas in mathematics.

Functions emerged as a way of stating the interrelationship of quantities at about the same time calculus came into being — and it's not surprising. Many of the concepts of calculus would be difficult to express without functional notation. Euler introduced writing functions as $f(x)$ in the 18th century, and we've been doing so ever since.

Once the Cartesian plane was linked to the graph of a function $y = f(x)$, algebraic expressions became visual. It was a fair exchange, the Cartesian plane could be used to present algebraic proofs of geometrical theorems, but it could also be used to give geometric representation to algebraic ideas. While it is possible to discuss such concepts as maximum and minimum values of a function by using a table of values to display the function, how on Earth could we use tables to discuss what is meant by bending up or bending down, to say nothing of points of inflection? Yes, it is possible to give a functional definition of bending down, on the order of $f(tx + (1-t)y) \leq tf(x) + (1-t)f(y)$ for $0 \leq t \leq 1$ — but even to an experienced mathematician, those are just symbols. Plot the graph of the function, and it's a whole lot easier to explain and see — and also verify with the tools of calculus.

Perhaps even more importantly, the Cartesian plane, coupled with the graph of $y = f(x)$, makes it possible to understand and analyze a much larger family of curves than contemplated by the Greeks. The Greeks, of course, were well aware of lines and circles, and they were also familiar with the family of curves that are still called conic sections — the parabola, the ellipse, and the hyperbola.

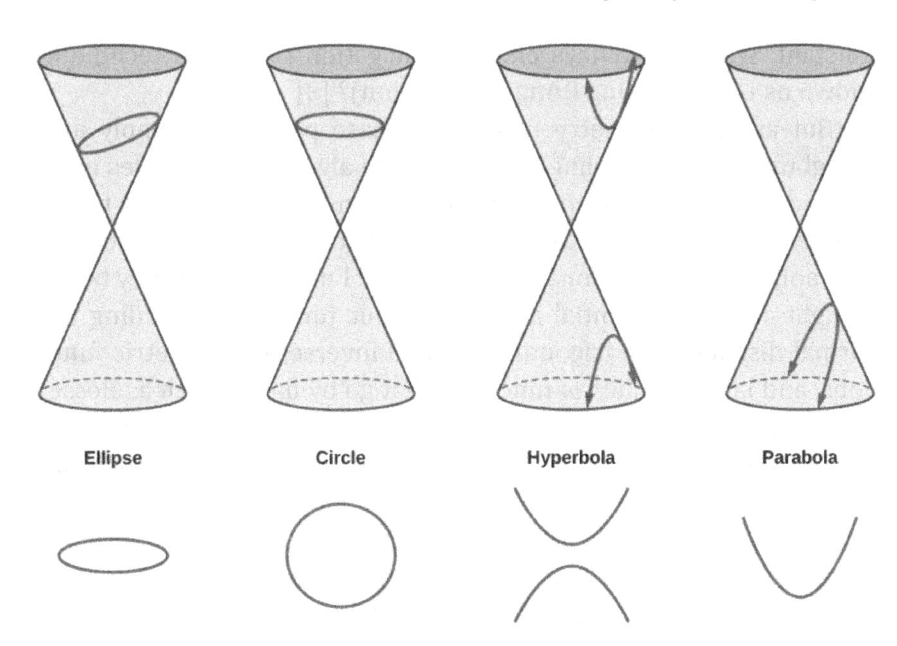

All of these curves were viewed by the Greeks as the result of intersecting a plane with a cone, as can be seen above. Different angles of intersections give different curves as pictured above. But how do you use them? Granted, the Greeks didn't contemplate the use of the parabola in quadratic optimization problems, such as finding the rectangular pasture with the largest area that can be enclosed using a given amount of fencing — but if they had, how would they have attacked the problem?

Also, the Greeks did not have the technology available that would have enabled them to use various properties of these curves. It certainly isn't clear that the Greeks knew that a line parallel to the axis of a parabola reflected off the tangent to the parabola would go through the focus of the parabola — and it's that property that enables our satellite dishes to receive transmissions. Did they know of a similar reflective property of an ellipse, which enables lithotripsy [1], a medical procedure for destroying kidney stones by concentrating radio waves at one of the foci? Were they aware that the difference of the distances from a point on a hyperbola to each of the two foci remains constant, which nowadays enables navigation through a technique known as LORAN (LOng RAnge Navigation)? [2]

But analytic geometry reveals all these properties simply and straightforwardly. Coupling geometry with algebra also enables us to use and visualize a large number of functions totally unknown to the Greeks. It isn't even clear, at least to me, that the Greeks even considered polynomials or rational functions. And I'm pretty sure they never thought about exponential and logarithmic functions (including the normal distribution), trigonometric and inverse trigonometric functions, and large families of functions that go by names such as Bessel functions, which are solutions to important types of equations.

There's more. The Cartesian plane enables us to graphically depict data of all sorts. Present information in tabular form, and it's hard to find patterns. Present the same data in an appropriately chosen Cartesian plane, and patterns quickly emerge. My favorite example of this is the Hertzsprung-Russell Diagram [3], which plots the luminosity vs temperature of stars.

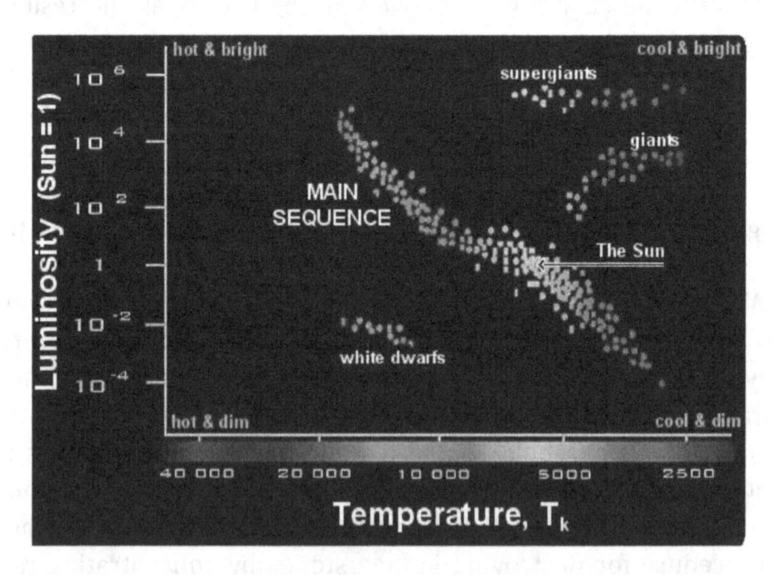

I'm not sure how stars were classified prior to the presentation of this diagram, but now we know there are four broad categories into which stars fall — and this diagram makes it clear what those categories are. This diagram is SO incredibly revealing that it is known as the Rosetta Stone of stellar astronomy.

Or take a time series — observations of a parameter of some sort plotted on the vertical axis against time on the horizontal axis. Again, present this information in tabular form and it may be difficult to discern a pattern. Present it in visual form on a Cartesian plane, and it might appear to be in the shape of an exponentially-damped sinusoidal function. Such functions are the solutions to a class of second-order differential equations which occur in many different environments, including springs in dashpots from mechanical engineering and RLC circuits from electrical engineering. The graph might suggest a mathematical model — which can then be tested against other datasets.

In the movie *Jerry Maguire*, Jerry expresses his love for Dorothy in a long-winded speech. Dorothy replies, "Just shut up. You had me at 'hello'." Well, I have some more remarks to make about the seductiveness of analytic geometry — but it had me at "makes geometry simpler."

Implicit Functions

The basic requirement for a function $y=f(x)$ is that it be single-valued; you can't give it one value of x and get two output values for y. If you define the square root of 4 to be a number which when squared equals 4, there are two possibilities; 2 and -2. So, in order to create a square root function, we have to pick one. We could define a function by choosing the positive square root for x values less than 100 and the negative square root for x values greater than or equal to 100, but we'd like the square root of a number to get larger as the number gets larger, so we choose the positive square root to be the square root function.

Now consider the circle $x^2+y^2=25$. This consists of all points at distance 5 from the origin; two of those points are (3, 4) and (−3, 4). So this circle isn't a function, but it can be constructed using two functions — the upper semi-circle function ($y \geq 0$) and the lower semi-circle function ($y \le 0$). Of course, we could construct this circle by using functions in many other ways.

Any expression of the form $g(x, y) = 0$, where g is a mathematical expression involving the letters x, y, and either functions or operations, is called an implicit function. To see why, let's look at one such expression.

$$\frac{y}{x^3} - \frac{x}{1+y^2} - 1 = 0$$

Yes, it would be possible to use the general solution to the cubic to solve for y in terms of x, or the general solution to the quartic to solve for x in terms of y. However, if you were to substitute $\sin x$ for one of the appearances of x, and $\ln y$ for one of the appearances of y, I can guarantee you couldn't do that. But, thanks to modern online graphing calculators, with a nod to Descartes, who made all this possible, we can get a look at it.

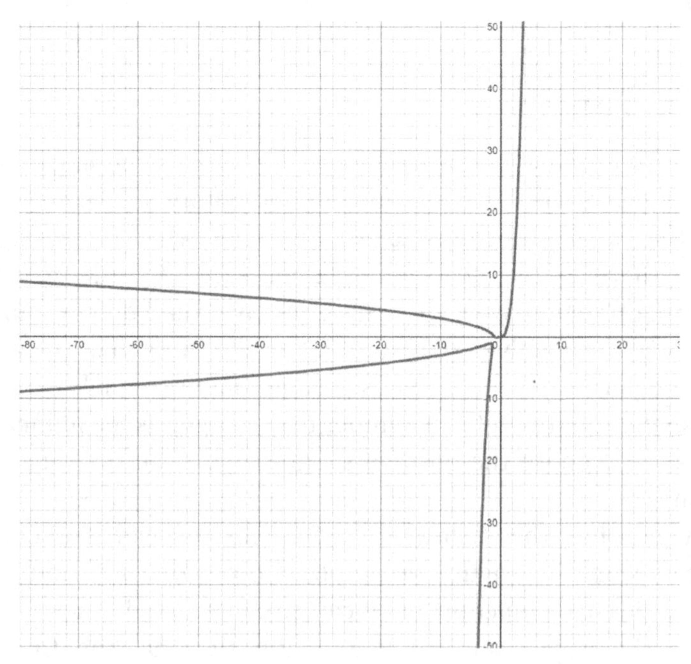

You might notice that this graph appears to be constructed of three functions — the upper branch (the one that's mostly above the x-axis), the mostly horizontal branch and the mostly vertical branch that's below the x-axis. But the important point is that the graph gives us some idea of the behavior of this particular function — and thus of any process modeled by the above equation.

Now, looking at this particular graph — and considering I just made up the equation off the top of my head — I don't think it's really likely that there's such a process. But who knows? There might be one out there somewhere, it's a big Universe.

And that brings us to one of the really seductive aspects of math — at least for me. Steven Weinberg, the Nobel-Prize winning theoretical physicist who just recently passed away, put it extremely well. He found it amazing that we can make these squiggles with paper and pencil that describe perfectly what is happening in the real world — at least, processes that are described by measurable parameters.

When I was about 10 years old, there was a partial eclipse of the Sun. It occurred in late afternoon, and the morning paper described exactly what was going to happen and when it was going to happen. I asked my father how people knew this, and he said that scientists were able to use mathematics to tell them this. I was just as amazed then as Steven Weinberg was — and I still am.

The Moving Dot

But I think the full beauty and utility of the Cartesian plane is only apparent when we consider parametric equations.

Although a picture may be worth a thousand words, a video is at least a short story. A graph of a function or an implicit function is a picture, but a graph of parametric equations is a video.

Of course, it is possible to make a graph a video in a somewhat limited sort of way. Think of the graph $y = f(t)$ as being presented on a video screen. At each time t, a dot appears at location $(t,f(t))$ in the Cartesian plane. As the clock t moves forward in time, the dot traces out the function.

For many functions, this is more than satisfactory. But it suffers from two limitations. The first is that the dot is tracing out a function, so the video has the dot moving from left to right on the screen. In addition, the dot's horizontal velocity is constant, so even though it may go uphill slowly (or quickly), only the velocity in the y-direction is capable of change. And it actually looks a little weird, because it goes up a steep hill quickly but up a gentle slope more slowly.

Here's how parametric equations work in the Cartesian plane. Once again, we will use a parameter t to act as a clock — but we now have two functions $f(t)$ and $g(t)$. We use these two functions to determine both the x and y coordinates of the same point at time t. That is, at any time t, $x = f(t)$ and $y = g(t)$. Just to be clear, if $x = 2t$ and $y = 3t^2 + 1$, at time $t = 1$ the moving point is located at (2, 4).

Because when math is fun, it's very seductive — and the grapher referenced in the next sentence is an immensely enjoyable app. Thanks to Dr. Adrian Jannetta (4) for making it available.

You might want to jump on this website and follow along as I use it. I may not be taking full advantage of the app's capabilities, apologies to Dr. Jannetta if so. What I'm going to do is show how to play different videos using the same graph.

The graph I'm going to use is the unit circle. I discussed earlier that this isn't a function, but parametric equations can be used to display graphs that are not necessarily function. So let

$$x(t) = \cos t \;\; y(t) = \sin t \;\; t_{max} = 6.28 \;\; t_{min} = 0$$

Position the big dot on the slider as far to the left as it will go, you will see a dot on the graph at (1,0) — the point on the curve at time $t = t_{min} = 0$. Now move the slider at a uniform rate from as far left as possible to as far right as possible. You will see the dot move on the curve at a uniform rate, traversing one full revolution in the counterclockwise direction.

Now change this slightly

$$x(t) = \cos(t^2) \;\; y(t) = \sin(t^2) \;\; t_{max} = 6.28 \;\; t_{min} = 0$$

Even though you are presumably moving the slider at a uniform rate, denoting a uniform passage of time, the dot speeds up as t increases, and it's really moving as you get closer to $6.28 = 2\pi$.

Try this one. Just for fun — and if I were teaching a class, this is what I'd do — see if you can predict what happens as you move the slider at a uniform rate from left to right.

$$x(t) = \cos(6.28t/(1+t)) \quad y(t) = \sin(6.28t/(1+t)) \quad t_{max} = 100 \quad t_{min} = 0$$

Incidentally, I noticed as I was doing this that if you watch this on a large screen (I'm a desktop guy), the slider dot has a solid black center and a light gray halo, if you click very slightly to the right of the halo the clock advances by a constant amount — in this case, tenths. Feel free to experiment with it.

You can modify this to make the dot move around the circle in a huge variety of different ways.

That's just working with known curves. Try making up one that has probably never been seen before and see what happens.

Lest you think this is just fun and games, parametric equations are an incredible tool to study the evolution of processes. When you see the track of a storm superimposed on a map, or the path of a rocket exploring the outer edges of the solar system, you're looking at parametric equations in the Cartesian planes.

But there's more.

Foxes and Rabbits — The Lotka–Volterra Equations

Remember those pairs of rabbits that Fibonacci hypothesized? Happily reproducing away, not a care in the world. Well, that makes for an interesting mathematical problem, but it's not indicative of what happens in the real world. If those rabbits were able to reproduce as Fibonacci hypothesized, the world would soon be overrun by rabbits. In the real world, there are foxes — which prey on the rabbits.

The Lotka–Volterra equations [5] describe the interaction of the populations of rabbits and foxes. These equations are differential equations, which one sees in calculus, but they are built on the following assumptions.

(1) The growth rate of rabbits is directly proportional to the number of rabbits. This certainly makes sense, twice as many rabbits will create twice as many bunnies. This growth rate incorporates the natural death rate of the rabbits.

(2) The decrease in the number of rabbits is proportional to the product of the number of foxes times the number of rabbits. This is because twice as many foxes will catch twice as many rabbits, and If there are twice as many rabbits available, a single fox will catch twice as many rabbits. Obviously, this is a simplification — but most models are originally presented as simplifications and are then modified.

(3) The growth rate of the foxes is proportional to the total number of rabbits they eat.

(4) The death rate of foxes is proportional to the number of foxes.

So how do parametric equations enter into this? Let $x(t)$ be the number of rabbits at time t, and let $y(t)$ be the number of foxes at time t. Obviously, there are a tremendous number of possible outcomes of this system — too many foxes, and both populations will die off; too many rabbits and, well, you can guess what happens. But with the right combination of initial populations and proportionality constants, the system is stable. As the rabbit population increases, so does the fox population, until the excess of foxes kill off so many rabbits that both populations decline. Then rabbit reproduction comes to the rescue, and the rabbits reproduce so rapidly that the fox population can grow again, and the cycle repeats.

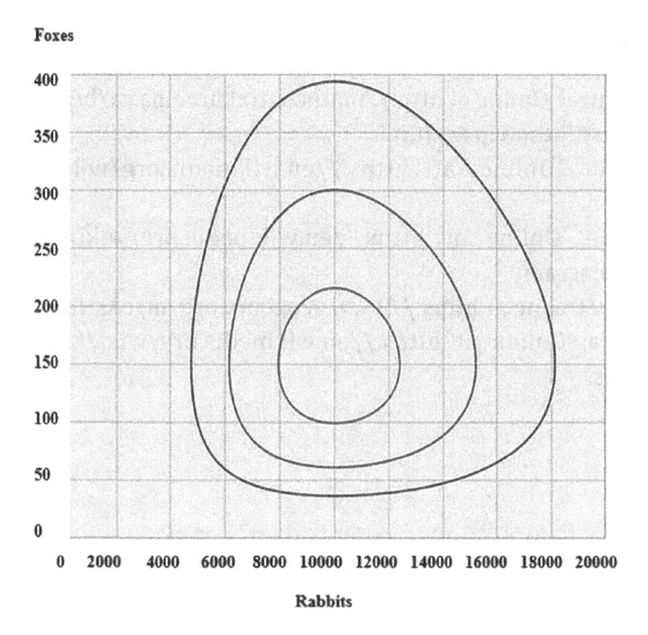

There are three different curves in this particular diagram. For instance, we can think of the outer curve as having started from initial populations of 14,000 rabbit and 50 foxes. As indicated in the previous paragraph, the system evolves in cyclic fashion — if a moving dot were placed on this particular curve, as time passes the moving dot moves in counterclockwise fashion. The other two curves represent other initial populations.

Remember the predictions that took place during the pandemic of how the disease would spread? Some of those predictions were the result of a similar system of differential equations. And, as time goes on, we find more and more ways to apply mathematics to the real world.

And in doing so, we can know the future — and hopefully change it for the better.

Bibliography

[1] Math Central. Online at http://mathcentral.uregina.ca/beyond/articles/Lithotripsy/lithotripsy1.html.
[2] Wikipedia. Online at https://en.wikipedia.org/wiki/Hyperbolic_navigation.
[3] Wikipedia. Online at https://en.wikipedia.org/wiki/Hertzsprung-Russell_diagram.
[4] Geogebra. Online at https://www.geogebra.org/m/cAsHbXEU.
[5] Wikipedia. Online at https://en.wikipedia.org/wiki/Lotka-Volterra_equations.

Chapter 5

Seduced by Mathematical Induction (and How to Avoid It)

I can remember the first interesting fact I discovered about numbers. It was occasioned by reading about the invention of the game of chess in Gamow's classic book *One, Two, Three ... Infinity* [1]. Gamow said that the inventor asked the King of Persia for one grain of wheat on the first square of the chessboard, two on the second, four on the third, and doubling the amount of wheat on every subsequent square — and in so doing he had asked for more wheat than had ever been grown on Earth. So I decided to try adding up the numbers of grains of wheat, and produced the following table.

Square	Number of Grains on Square	Total
1	1	1
2	2	3
3	4	7
4	8	15
5	16	

At this point I stopped, because I had noticed something very striking — the total in the last column was always 1 less than the next number in the second column. I did a few more rows, and it always worked out that way. I'm pretty sure that was as far as I took it at the time, but I do recall wondering if this was always true, and if it was always true, how would you ever know? You would have to establish an infinite number of facts, and how could you possibly do that? Math has an answer.

The Principle of Mathematical Induction

Well, *mathematicians* have an answer; they have established a technique for establishing an infinite number of facts — and it's very clever and very powerful. It's known as the Principle of Mathematical Induction, or PMI, and is usually first introduced in Algebra II or Precalculus. It works like this: Assume that $P(n)$ is some proposition involving the positive integer n; this is a short way to describe an infinite number of propositions, one for each positive integer. Suppose we know two things; first, that $P(1)$ is true, and second, that if we assume $P(n)$ is true, we can deduce that $P(n + 1)$ is true. The second fact is usually stated "$P(n)$ implies $P(n + 1)$," for which the mathematical shorthand is $P(n) => P(n + 1)$. Incidentally, I've always found mathematical shorthand seductive, as it aids in the mathematical aesthetic of producing truths with the fewest possible number of symbols. Perhaps that's some of what Millay had in mind when she talked about "Beauty bare."

Anyway, it's easy to see why the Principle of Mathematical Induction works. We know that any non-empty collection of positive integers has a least positive integer in it. Suppose that we have a collection of propositions $P(n)$, one for each positive integer n, that satisfy the two facts in the previous paragraph. Suppose that S is the set of all positive integers n for which $P(n)$ is false. If S is non-empty, then it has a least positive integer N. So $P(N)$ is false. But we assume $P(1)$ is true, and since $P(n) => P(n + 1)$ for all n, if we let $n = 1$ we have $P(1) => P(2)$, so $P(2)$ is true. Applying $P(n) => P(n + 1)$ with $n = 2$, we have $P(2) => P(3)$, and so $P(3)$ is true. Proceeding up the ladder, we eventually conclude that $P(N)$ is true. However, we know $P(N)$ is false, and since a proposition cannot be simultaneously true and false, the set S must be empty. Therefore, $P(n)$ is true for all n.

The Principle of Mathematical Induction is one of the most powerful tools in the mathematical toolkit — and frankly, the first time I saw it, I failed to appreciate it. It first came up in an Algebra II class at the end of the semester — and who remembers the stuff they teach you at the end of the semester? But I became a PMI addict when I started working towards my doctoral thesis.

It's funny how things work out. While I was an undergraduate, my worst experience in mathematics came in a course called "mathematical analysis." It was taught by Shizuo Kakutani, a brilliant mathematician, from a book called *Principles of Mathematical Analysis*, by Walter Rudin [2]. Rudin's book is incomprehensible the first time you read it — or try to. There isn't a single picture in the entire book, and Rudin's sense of mathematical elegance is to prove a theorem using the least amount of characters. The second time you read it, you have a sense of what mathematical analysis is trying to accomplish. The third time you read it, it's a classic.

But the first time I read it, it was a disaster. I got a 76 the first semester and upped it to an 80 the second semester — and spent the remainder of my undergraduate career avoiding anything that remotely smacked of mathematical analysis. When I went to Berkeley for graduate studies, a course in mathematical analysis was standard for first-year students, and I was fortunate enough to have Bill Bade as the instructor. All of a sudden, analysis became comprehensible — perhaps the only epiphany I've ever experienced. When I passed my oral comprehensives, I ran into Bade in the hallway, and practically tackled him (not an easy task when the tackler is 5′9″ and weighs about 110 and the tacklee is a 6′2″ 200 pounder). I recall saying something like, "Please take me as a thesis student, if you don't I'm probably going to end up running a dry-cleaning establishment in the Bronx." Anyway, as fate would have it, he specialized in analysis, and in reading one of his papers, there was an absolute honey of a PMI proof. It was just gorgeous. I was hooked. So much so that most of the papers I wrote during the first half of my research career featured variations on the theme I found in his paper. So I have warm and fuzzy feelings about PMI.

And PMI can be used to establish the conjecture that first struck me that the sum of the first n powers of 2 is one less than the next power of 2. This is often used as one of the early examples of the power of PMI in the courses where it is taught.

We start by stating what $P(n)$ is. In this case, we write $P(n)$: $1 + 2 + \cdots + 2^n = 2^{n+1} - 1$. The truth of $P(1)$ is easy to check, because the left-hand side of $P(1)$ is $1 + 2^1 = 3$, and the right-hand side of P(1) is $2^{1+1} - 1 = 3$.

Now comes what, for most students, is the hard part. The first is actually writing out what $P(n + 1)$ says, I generally tell students just to substitute $n + 1$ wherever they see an n in the expression for $P(n)$. We now have to show that $P(n) \Rightarrow P(n + 1)$, which means that we have to show that $P(n + 1)$ is true if we know that $P(n)$ is true. Making the substitution of $n + 1$ for n, we see that $P(n + 1)$: $1 + 2 + \cdots + 2^{n+1} = 2^{(n+1)+1} - 1$; the term on the right is $2^{n+2} - 1$. Starting off with the left-hand side of $P(n + 1)$, doing a little algebra produces the desired result.

$$1 + 2 + \cdots + 2^{n+1} \qquad \text{writing in the penultimate term}$$
$$= (1 + 2 + \cdots + 2^n) + 2^{n+1}$$
$$= (2^{n+1} - 1) + 2^{n+1} \qquad \text{using the assumption that } P(n) \text{ is true}$$
$$= 2 \times 2^{n+1} - 1 = 2^{n+2} - 1 \qquad \text{and we're done}$$

There's absolutely nothing wrong with this; it supplies a good introduction to the use of PMI — but there's a seductive way to prove this without PMI.

March Madness, Classic Version

Back when the world was young and they didn't have those play-in games, March Madness, the NCAA Championship basketball tournament, was a standard single-elimination tournament with $64 = 2^6$ entrants. The first round consisted of $32 = 2^5$ games, the 32 winners would then play $16 = 2^4$ games, and so on until the final. The total number of games was 1 final, 2 semi-finals, 4 quarterfinals, 8 round of 16 matches, 16 second round matches and 32 first round. The total number of games was therefore the sum

$$1 + 2^1 + 2^2 + 2^3 + 2^4 + 2^5$$

But each game produces a loser, and the tournament continues until there is only 1 team that hasn't lost a game — the tournament winner. So the number of entrants – $64 = 2^6$ — is the sum of the number of losers + 1 (the winner). But the number of games equals the number of losers, and that's just the number of entrants minus 1.

This really isn't a formal proof — a rigorous proof would require dotting some i's and crossing some t's. In particular, we only did this for a 64 team tournament, but who cares? It's more than enough to see that the argument is true for all possible sums of powers of 2.

It's even possible to take this a little further. Suppose we wanted to compute the sum of powers of 5. Imagine a tournament in which there are 5^N entrants, grouped into 5^{N-1} mini-tournaments of 5 teams each. Each mini-tournament produces a winner and 4 losers. The winners are then grouped into more 5 team mini-tournaments, and the whole tournament continues in this fashion until the last mini-tournament of 5, from which the winner emerges.

Once again, the number of entrants, 5^N, is equal to the sum of the number of losers plus 1. In the final, there is 1 winner and 4 losers. There are 5 mini-tournaments in the penultimate round, again producing 4 losers. The total number of mini-tournaments is therefore $1 + 5 + 5^2 + \cdots + 5^{N-1}$, and so the total number of losers is 4 times this number. Adding the 1 winner to the losers, using the idea that number of entrants = number of losers + 1, we see that

$$5^N = 4 \times (1 + 5 + 5^2 + \cdots + 5^{N-1}) + 1$$

And so

$$(1 + 5 + 5^2 + \cdots + 5^{N-1}) = (5^N - 1)/4$$

But why stop there? If our mini-tournaments are all of size r and each mini-tournament produces 1 winner, we see that

$$(1 + r + r^2 + \cdots + r^{N-1}) = (r^N - 1)/(r - 1)$$

It really doesn't matter that we looked at r as a positive integer, the argument is the same for any real number — and any complex number as well (and we'll get to those later on in the book). True, it's a little hard to envision what a tournament with $\pi - 4.38i$ entrants looks like, but as long as only 1 winner emerges from each group of $\pi - 4.38i$ contestants, the above argument works just fine.

The GOAT — Candidate #1

If you follow any sport, there's almost certainly likely to be an argument as to who is the GOAT — the Greatest Of All Time. In men's swimming, you can be pretty sure it's Michael Phelps, but you'd get an argument as to who is the GOAT in tennis, or golf, or practically any other sport.

And you might have a similar argument in mathematics, but one guy who's guaranteed to be in the GOAT conversation is Carl Friedrich Gauss [3]. Gauss was first noticed by an elementary school teacher who felt the need for a break during an arithmetic lesson and instructed the students to take their slates and add up the numbers from 1 to 100 in the interim. When the teacher returned from his break, all but one of the students were still busy; Gauss had written the correct answer, 5,050, on his slate. When the astonished teacher inquired how Gauss had managed to get the correct answer in such short order, the child showed him what has come to be known as the Gauss trick.

Gauss wrote the numbers in ascending order (leaving some out), and below it the numbers in descending order. It looked like this

$$\begin{array}{ccccccccccccc} 1 & + & 2 & + & 3 & + & \cdots & + & 98 & + & 99 & + & 100 \\ 100 & + & 99 & + & 98 & + & \cdots & + & 3 & + & 2 & + & 1 \end{array}$$

Gauss noticed that all the column sums, $1 + 100$, $2 + 99$, etc., totaled 101. There were 100 such columns, so the sum of both rows was $100 \times 101 = 10{,}100$. Because there were two rows with the same sum, the sum of the numbers from 1 to 100 was therefore half of 10,100, or 5,050.

When you're the GOAT, you just do things like this, even when you're just a kid. (We math types like plays on words as well as plays on numbers.) And you don't have to be the GOAT to realize that the same trick shows that the sum of the integers from 1 to n is $n(n+1)/2$, because if you do the same thing for the numbers from 1 to n, there will be n column pairs, each with a sum of $n+1$.

And it shouldn't surprise you to know that there's a fairly simple proof of this using PMI. If $P(n)$ is the proposition that $1+2+\cdots+n = n(n+1)/2$, $P(1)$ is trivially true, as both sides are equal to 1. Assuming that $P(n)$ is true, $P(n+1)$ would be the proposition that $1+2+\cdots+n+1 = (n+1)((n+1)+1)/2 = (n+1)(n+2)/2$, once again substituting $n+1$ any place we see an n in $P(n)$. But, very similar to the proof for $P(n) \Rightarrow P(n+1)$ for the powers of 2, we have

$1+2+\cdots+n+1$ writing in the penultimate term
$= (1+2+\cdots+n)+n+1$

$= n(n+1)/2+(n+1)$ using the assumption that $P(n)$ is true

$= (n+1)(n/2+1) = (n+1)((n+2)/2)$ and we're done

Neat and tidy — but nowhere near as appealing as the Gauss trick. But we're not done with this particular result. Check out the following diagram

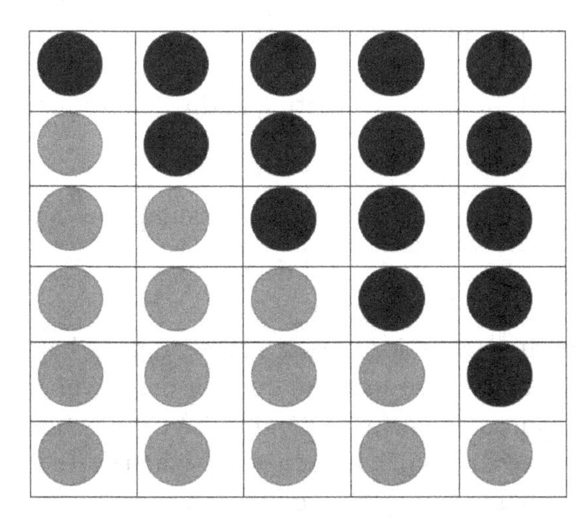

This is a picture for the sum of the integers from 1 through 5. There are $1 + 2 + 3 + 4 + 5$ dark dots and $1 + 2 + 3 + 4 + 5$ light dots, in a total of 6 rows and 5 columns = 30 dots. So the sum of the integers from 1 through 5 is $\frac{1}{2} \times 5 \times 6$ — and the picture would be the same for the sum of the first n integers, although with more dots.

If you look a little closer, this is just a visual of the Gauss trick! The first column is 1 dark dot plus 5 light dots, a total of 6 dots. The second column is 2 dark dots plus 4 light dots, again a total of 6 dots. And, of course, the last column is 5 dark dots plus 1 light dot, a total of 6 dots.

The Sum of Odd Integers

There are a LOT of problems that can be solved using PMI — like I said, it's one of the most powerful tools in the kit — and I haven't managed to kick my PMI addiction to this day. Here's another one that students practice on when first exposed to PMI. If you look at the sums of the successive odd integers, they always total a square, such as $1 + 3 + 5 + 7 = 16 = 4^2$. The PMI would be used to prove the proposition $P(n)$: $1 + 3 + \cdots + 2n - 1 = n^2$.

This is the easiest use of the PMI yet. $P(1)$ is obviously true, as $P(1)$ is just the proposition $1 = 1^2$. If we assume that $P(n)$ is true, then $P(n + 1)$ is $1 + 3 + \cdots + 2((n + 1)-1) = (n + 1)^2$. By now, the proof technique should be pretty familiar.

$$1 + 3 + \cdots + (2n + 1) \qquad \text{writing in the penultimate term}$$

$$= (1 + 2 + \cdots + 2n - 1) + (2n + 1)$$

$$= n^2 + 2n + 1 \qquad \text{using the assumption that } P(n) \text{ is true}$$

$$= (n + 1)^2 \qquad \text{and we're done}$$

But part of the aesthetic of math — that "Beauty bare" that Millay was talking about — is to use just enough firepower to get the job done. Using a tool as powerful as PMI on this proposition is like grabbing a sledgehammer to kill a fly. Especially when there's such an easy visual proof available simply by looking at a chessboard.

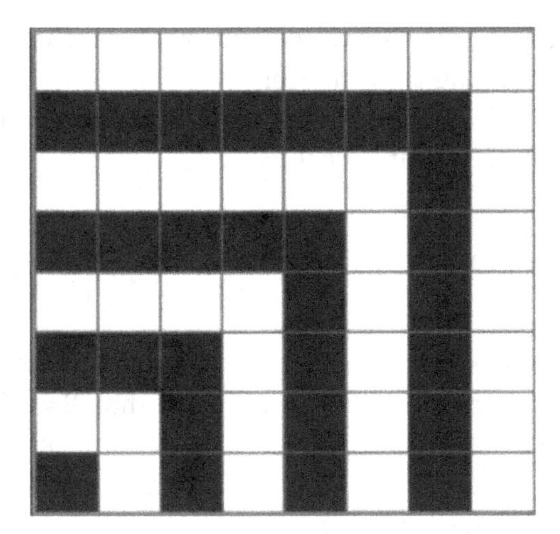

Proceeding up from the lower left corner, we see alternately outlined in black and white the odd integers $1 + 3 + 5 + 7 + 9 + 11 + 13 + 15 = 8^2$. You probably could have convinced yourself of the truth of this result simply by looking at the 5×5 rather than the 8×8.

The GOAT — Candidate #2

All those PMI proofs are setting students up for one of the first of the great algebraic results in mathematics — the Binomial Theorem. A general version of the Binomial Theorem for rational exponents was first proved by GOAT Candidate #2, Isaac Newton [4], when he was a student at Cambridge. The scuttlebutt surrounding this particular discovery is that Newton showed the proof to Isaac Barrow, the first holder of the Lucasian Chair of Mathematics at Cambridge — which was recently held by Stephen Hawking. Barrow looked at it and said something on the order of, "Well, kid, I was planning on stepping down anyway, and now is as good a time as any," and handed the Chair over to Newton. Events were to prove that the Chair was in good hands.

At any rate, the Binomial Theorem for exponents which are positive integers is a challenge worthy of the PMI. I really think that algebra students initially learn the PMI simply so they can prove the Binomial Theorem. It's a great theorem, and had Pythagoras discovered it, there probably would have been another barbecue. It's easily worth sacrificing a dozen oxen.

Even with the PMI, though, most algebra students find the proof somewhat daunting — but here it is.

$$P(n) \text{ is the proposition that } (x+y)^n = \sum_{k=0}^{n} \frac{n!}{k!(n-k)!} x^k y^{n-k}$$

In order to work through the proof, students need to know that $n! = 1 \times 2 \times \cdots \times n$ if n is a positive integer, and that $0! = 1$, which is also a definition, because it makes formulas work out nicely.

As usual, establishing $P(1)$ is pretty straightforward. Both the left and right sides are easily seen to be $x + y$. Also as usual, the devil is in the details of the induction step $P(n) \Rightarrow P(n+1)$, and this time we have our work cut out for us.

Because the proof has a number of steps that require lengthy expressions, I'm going to number each step and give the reasons for each step after the proof is completed, sort of like a delayed two-column proof.

$$(x+y)^{n+1} = (x+y)(x+y)^n = (x+y)\sum_{k=0}^{n} \frac{n!}{k!(n-k)!} x^k y^{n-k} \quad (1)$$

$$= \sum_{j=0}^{n} \frac{n!}{j!(n-j)!} x^{j+1} y^{n-j} + \sum_{k=0}^{n} \frac{n!}{k!(n-k)!} x^k y^{n+1-k} \quad (2)$$

$$= \sum_{k=1}^{n+1} \frac{n!}{(k-1)!(n+1-k)!} x^k y^{n+1-k} + \sum_{k=0}^{n} \frac{n!}{k!(n-k)!} x^k y^{n+1-k} \quad (3)$$

Step (1) is simply using the assumption that $P(n)$ is true, substituting the complicated summation for $(x+y)^n$.

Step (2) expands the last expression on the right of (1) using the distributive law, but uses the letter j as the index of summation for the first sum. The reason for doing this is that by substituting $j = k - 1$ in the first sum in (2), we get the first sum in (3). Notice that $k = j + 1$, so that when $j = 0$, $k = 1$, and when $j = n$, $k = n + 1$.

If we take a look at (3), there are several things worth noticing. Both of these are sums, using k as an index, of the term $x^k y^{n+1-k}$, so we can collect terms for the common values of k in the two sums. Those common values are $k = 1, 2, \ldots, n$. When $k = n + 1$, there is no term in the second sum, but there is one term in the first sum. That term is

$$\frac{n!}{n!0!}x^{n+1}y^0 = x^{n+1} = \frac{(n+1)!}{(n+1)!0!}x^{n+1}y^0$$

When $k = 0$, there is no term in the first sum, but there is one term in the second sum. That term is

$$\frac{n!}{0!n!}x^0y^{n+1} = y^{n+1} = \frac{(n+1)!}{0!(n+1)!}x^0y^{n+1}$$

I've written the right-hand term in each of the above two equalities in order to see that these terms "fit in" nicely in the final summation.

For $k = 1, 2, \ldots, n$ we can add the coefficients of $x^k y^{n+1-k}$ in each of the two sums in (3). Adding these coefficients, we obtain the following expression for the coefficient of $x^k y^{n+1-k}$.

$$\frac{n!}{(k-1)!(n+1-k)!} + \frac{n!}{k!(n-k)!} = \frac{n!}{(k-1)!(n-k)!}\left(\frac{1}{n+1-k} + \frac{1}{k}\right)$$
$$= \frac{n!}{(k-1)!(n-k)!}\left(\frac{n+1}{k(n+1-k)}\right) = \frac{(n+1)!}{k!(n+1-k)!}$$

So we've now seen that for $k = 1, 2, \ldots, n$ and the two previous cases $k = 0$ and $k = n + 1$, the coefficient of $x^k y^{n+1-k}$ is the last expression on the right hand side of the last equation. Consequently,

$$(x+y)^{n+1} = \sum_{k=0}^{n+1} \frac{(n+1)!}{k!(n+1-k)!}x^k y^{n+1-k}$$

This is the expression we need to establish to prove $P(n+1)$, and by PMI we've proved the Binomial Theorem. I'm guessing neither Edna nor Euclid would see a whole lot of "Beauty bare" in that proof, but it has the virtue of getting the job done.

But there's a seductive way to prove this theorem. Those coefficients of $x^k y^{n-k}$ are known as binomial coefficients, but they are sometimes written $C(n, k)$, because they also crop up in combinatorics as the number of combinations of n things taken k at a time. Simply put, a number such as $C(52,5)$ can be viewed as the number of different lottery tickets that pick 5 numbers from the numbers 1 through 52. If you buy a lottery ticket with the numbers 1, 3, 4, 9 and 15, it doesn't matter in what order they draw these numbers out of the hat (or wherever they draw them from), or in what order they're printed on the ticket, it's the same lottery ticket. For those of us who used to play poker back in the day when 5-card draw was the standard way to play (rather than Texas Hold-'Em), $C(52,5)$ is the number of different 5 card poker hands. If your hand contains the Aces of spades, diamonds, and clubs, and the 5s of spades and hearts, it doesn't matter how you sort them, it's the same full house.

Once upon a time, the binomial coefficient $C(n, k) = n!/(k!(n-k)!)$ was demonstrated early in a course in combinatorics but nowadays it generally shows up in high school algebra. If you haven't seen it and you're not willing to trust me, it's easy to find online.

With that as a preamble, let's move on to a much more seductive proof of the Binomial Theorem. Let's take a look at an expression such as $(x+y)^2$, and multiply it out **without** collecting terms. We get

$$(x+y)^2 = (x+y)(x+y) = xx + xy + yx + yy$$

Notice that the expression on the right consists of all 2-letter words that can be made from the letters x and y. Now let's do the same for $(x+y)^3$. Again, we get

$$\begin{aligned}(x+y)^3 &= (x+y)(x+y)(x+y) \\ &= xxx + xxy + xyx + xyy + yxx + yxy + yyx + yyy\end{aligned}$$

And again, this is all three letter words that can be made from the letters x and y. This isn't at all surprising, each word is constructed by multiplying one term each from the three factors $x+y$ whose product is $(x+y)^3$.

Notice that in the above expression, there are three words which have the same value x^2y, these words are *xxy*, *xyx*, and *yxx*. So the coefficient of x^2y in the expansion of $(x+y)^3$ is simply the number of three-letter words which consist of 2 *x*'s and 1 *y*.

Therefore, if we expand $(x+y)^n$ and want to know the coefficient of x^ky^{n-k}, we need merely compute the number of *n*-letter words which have precisely *k x*'s and *n–k y*'s. But this is simply $C(n,k)$! To see this, let's just take a simple example, the number of 6 letter words with 2 *x*'s and 4 *y*'s. We take a lottery ticket containing two numbers from 1 through 6; let's say those numbers are 3 and 5. We stick *x*'s in positions 3 and 5, and *y*'s in the remaining slots. This would create the word *yyxyxy*. The number of such words are $C(6,2)$, the number of lottery tickets with 2 numbers on them chosen from the numbers 1 through 6.

And that's the proof of the Binomial Theorem.

The PMI proofs in this chapter are a forerunner to some of the ingenious ways it can be used. If you take a course in calculus, one of the first rules you encounter is that the derivative of the function x^n is nx^{n-1}. The standard way of proving this — and every math textbook feels obliged to prove every single statement it makes, which adds not only to reading difficulty but also to acute boredom — is to use limits and the Binomial Theorem. However, it's a really easy proof using PMI. De Moivre's Theorem on the powers of complex numbers is also a setup for PMI.

It's an extremely useful tool, and I find it seductive simply to know that there's a powerful technique for proving an infinite number of separate propositions in case I am called upon to do so. But I hope I've been able to show in this chapter that there are cases in which simple flowers can be more seductive than elegant bouquets.

Bibliography

[1] G. Gamow, *One Two Three ... Infinity*. New York NY: Viking Press, 1961.
[2] W. Rudin, *Principles of Mathematical Analysis*. International Series of Pure and Applied Mathematics, The Netherlands: Elsevier, 1976.
[3] MacTutor. Online at https://mathshistory.st-andrews.ac.uk/Biographies/Gauss/.
[4] MacTutor. Online at https://mathshistory.st-andrews.ac.uk/Biographies/Newton/.

Chapter 6

Seduced by Calculus I

I was always attracted to math — but I was seduced by calculus.

Although the term "pre-calculus" is used to describe the course — in high school or college — that precedes and supplies the needed prerequisites for calculus, I've always thought that "pre-calculus" really describes the Dark Ages when we were groping for an understanding of the Universe. Calculus is not just a mathematical *tour de force*, it supplies critical tools for physics, engineering, chemistry, biology and business.

In his classic *The Two Cultures*, C. P. Snow discussed two concepts in science which parallel levels of knowledge in the humanities [1]. Understanding velocity and acceleration, Snow argued, was the equivalent of knowing how to read. Understanding the Second Law of Thermodynamics was the equivalent of having read a work of Shakespeare. Both require calculus — as do the great theories of motion and mechanics discovered by Newton, the electromagnetic theories of Maxwell, Einstein's Theory of Relativity, and Schrodinger's quantum mechanics. Calculus describes how organisms grow, how disease spreads — and at what price a business should sell its product to maximize its profits.

But beyond its utility, calculus is beautiful — not only in its approach to the study of how quantities change, which underpins all of calculus, but in the problems that always bothered me while I was taking the courses that preceded calculus. You'll see some of those questions in the next few chapters, in what context they arose, and how calculus does an incredible job of answering them.

Simply the Best

Generally, the first encounter one has with an optimization problem occurs in algebra, or maybe pre-calculus, when you learn about quadratic optimization. The basic example is usually something like this: a farmer has 200 meters of fencing with which to enclose a rectangular field. What is the largest area that he can enclose?

So you let L be the length of the field, W the width, and notice that the perimeter of the field, $2L + 2W$, satisfies $2L + 2W = 200$. The area of the field is LW, but if you solve the perimeter equation for L in terms of W, you see that $L = 100 - W$. If you substitute this for L, you see that $A = (100 - W)W = 100W - W^2$, which we write as $A = -W^2 + 100W = -(W^2 - 100W)$. You can complete the square of this quadratic and factor to see that $A = -(W^2 - 100W + 2{,}500) + 2{,}500 = 2{,}500 - (W - 50)^2$. Since squares are never negative, the largest value of A occurs when $W = 50$ meters, and $A = 2{,}500$ square meters.

OK, this technique is known as quadratic optimization. It's pretty cute, and can be used for a lot of problems, but it's limited, not surprisingly, to quadratics or variations thereof. And perhaps the first glimpse of the practical power of calculus comes when one realizes its ability to do any type of single-variable optimization problem — the type you see in first semester calculus. This involves trying to find the best way (maybe the fastest, maybe the cheapest) to do something. Calculus does this so well that there's a natural tendency to pursue this approach whenever one is confronted with an optimization problem. And that's what I did when I was given the following problem during an interview for a summer job with an engineering firm while I was still in college.

So here's the problem. You have a factory 20 miles from a river, and a power station 10 miles from the same side of that river. The nearest point on the river from the factory is located 50 miles from the nearest point on the river from the power station. You need to build a power line from the power station to the river to the factory. What is the shortest possible length of the power line?

Piece of cake, I thought. I drew the following diagram and went to work.

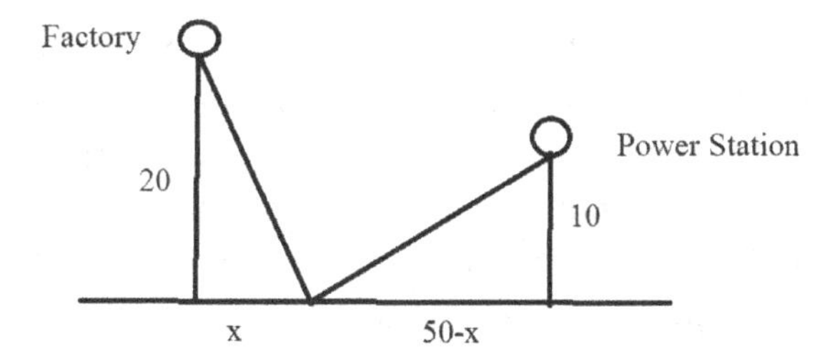

I quickly recognized as you no doubt will as well, that the total length L of the pipe was the sum of the hypotenuses of two right triangles. I duly wrote

$$L=\sqrt{x^2+400}+\sqrt{(50-x)^2+100}$$

Trust me, you don't want to see the gory details of what followed, which involved a lot of algebra in which I took the derivative of this function, set it equal to zero, and solved for x — standard operating procedure for optimization problems. What you DO want to see is the 15-second solution to this problem.

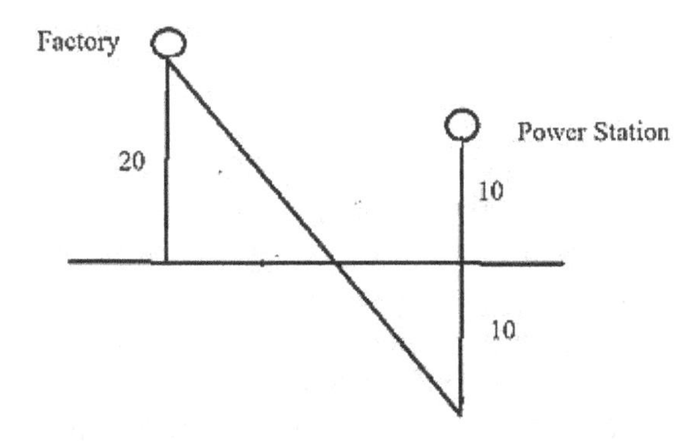

A straight line is the shortest distance between two points!

An amazingly cute solution to the problem — but it's hard to see this as anything that simplifies optimization problems in general; it was just something that worked for one particular problem (although you can see a variation of it in the proof of Snell's Law in optics).

Then, some years later, I found myself teaching Calc I, and the book had the following standard optimization problem — find the area of the largest rectangle that can be inscribed in a circle of radius R. The standard way to do this is to let the length of the rectangle be L, the width W, and to notice that the length and the width form a right triangle whose hypotenuse is the diameter of the circle. Consequently, the problem reduces to maximizing LW subject to the constraint $L^2 + W^2 = 4R^2$. So you solve for L in terms of W (or W in terms of L), substitute to reduce $A = LW = W(4R^2 - W^2)^{1/2}$ (as we always tell students, you can differentiate powers, but you can't differentiate roots), take the derivative, set it equal to 0, and solve. Not as grim as the factory pipeline problem, but still, you have plenty of chances to make an error.

You can discover a lot of math just by fooling around. I don't know what persuaded me to try this particular approach, but instead of using L and W as variables, I let ϑ be the angle between the diagonal of the rectangle (the diameter of the circle) and L. Then the length $L = 2R \cos \vartheta$, the length $W = 2R \sin \vartheta$, and the area of the rectangle is $4R^2 \sin \vartheta \cos \vartheta$. But, from trig, $\sin 2\vartheta = 2 \sin \vartheta \cos \vartheta$, and thus the area of the rectangle is $2R^2 \sin 2\vartheta$. Look Ma, no calculus! The function $\sin 2\vartheta$ has a maximum of 1 when ϑ equals 45°, so the maximum area of the rectangle is $2R^2$.

The same approach works if one wants to find the maximum area of a rectangle inscribed in an ellipse. If we parameterize the ellipse as $x = a \cos t$ and $y = b \sin t$, then if (x, y) is the vertex of the rectangle in the first quadrant, the area of the rectangle is $(2x)(2y) = 4xy = 4ab \cos t \sin t = 2ab \sin 2t$, again maximized when $t = 45°$.

The idea of using trigonometric variables is a staple in second-semester calculus, as one of the most powerful techniques for integrating a function is known as trigonometric substitution. Trig provides a convenient shorthand for dealing with functions of the form $\sqrt{a^2 - x^2}$. These functions arise naturally in problems (such as the one above) involving the use of the Pythagorean Theorem and lead to equations which require the student to do lots of differentiation of expressions such as $\sqrt{a^2 - x^2}$. This is tedious and students often make errors differentiating these expressions. Moreover, it leads to having to solve equations with similar expressions, never a fun job. Using trigonometric variables not only makes life considerably simpler, in some instances — such as finding the area of the largest rectangle that can be inscribed in a given ellipse — it obviates the need for calculus altogether. Love may mean never having to say you're sorry, but trigonometric variables mean never having to say $\sqrt{a^2 - x^2}$.

Now, the use of trig variables in optimization problems isn't a miracle cure. There are some problems, such as the one above, which it absolutely obliterates. There are also some in which it doesn't appear to help much. But any time the hypotenuse of a triangle appears, it's worth looking at. What follows are a few of my favorites, I leave you to find your own. Some of the ones below are staples of a first-semester calculus course, some are ones I just decided to try to compare the two approaches (distances and trig variables).

Crossing a River

I'm going to present most of the subsequent problems in a general format — specifying parameters as letters (such as R) rather than numbers. In general, it doesn't make the problem harder, and sometimes it enables useful observations.

We're standing on one side of a river of width W. On the opposite shore, located a distance D downstream, is a desired destination. We can travel across the river at a rate r (by boat or swimming), and can travel on land at a rate s (by running or by a bicycle we carried in the boat). Where should we land on the opposite side in order to minimize the total travel time?

The rectilinear approach is to assume that we land at x on the other side, where $0 \leq x \leq d$. The total time T is given by $T =$ time on river + time on land $= (x^2 + W^2)^{1/2}/a + (d - x)/b$. Differentiate, set equal to zero and solve — and don't forget to check the endpoints.

Any time you have to differentiate roots, there is a chance for error. This problem is so much easier — and the solution much more insightful, if one lets ϑ be the angle between the perpendicular drawn from the starting point to the opposite shore, and the landing point. Then the total time T is given by $T = W \sec\vartheta/a + (d - W\tan\vartheta)/b$. Differentiating, and setting equal to 0, we obtain the equation

$$(W/a)\sec\vartheta\tan\vartheta - (W/b)\sec^2\vartheta = 0$$

This quickly reduces to $\tan\vartheta/a = \sec\vartheta/b$, and multiplying both sides by $a\cos\vartheta$ (which can never be 0, since we have to get to the other side), we see that $\sin\vartheta = a/b$, where $0 \le \vartheta \le \tan^{-1}(d/W)$.

This problem showcases one of the advantages of using trig variables. The calculus of differentiating trig functions is already built in — students learn the derivatives of all six trig functions relatively early in the game. But differentiating complicated expressions involving square roots offers many chances for error. Furthermore, after the differentiation has been completed, doing the necessary algebra offers its own challenges — and every time a challenge presents itself, there's a chance of error. Not only is differentiating T straightforward in this problem, solving for the critical value of ϑ presents no problem.

The next problem reveals another advantage of this approach.

The L-Shaped Corridor

I'm aware that some may be reading this with interest but are not yet sold. Most of us who encounter these problems are comfortable with distance as the variable. But I think I can close the deal with the following example.

It's another classic problem. You need to move a pipe around a corner connecting two long corridors of width a and b. What is the longest pipe you can move?

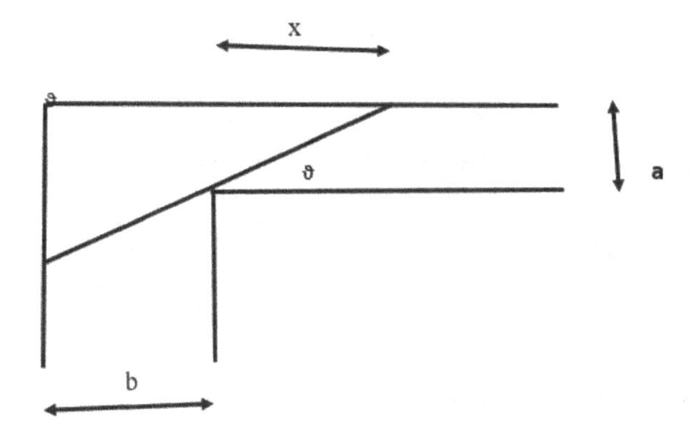

First, let's set it up in terms of x. Let L be the total length of the pipe, x the distance along the wall indicated in the diagram. Then the portion of the pipe from the wall in corridor a to the corner is $(x^2 + a^2)^{1/2}$. If u denotes the length of the pipe from the corner to the wall in corridor b, then $u = (x^2 + a^2)^{1/2}/x$ by similar triangles. So the total length of the pipe $L = (x^2 + a^2)^{1/2} + (b(x^2 + a^2)^{1/2}/x) = (x^2 + a^2)^{1/2}\,(1 + b/x)$. At this point I am going to play the "author's privilege" card and NOT carry this through to completion — but don't let me stop you from doing so.

But I'll finish WAY before you. If I use trig variables, the length of the pipe is $a \csc \vartheta + b \sec \vartheta$. Differentiating and setting the result equal to 0 yields $-a \csc \vartheta \cot \vartheta + b \sec \vartheta \tan \vartheta = 0$. Adding the first term to both sides and substituting sines and cosines where appropriate results in

$$a \cos \vartheta / \sin^2 \vartheta = \mathrm{b} \sin \vartheta / \cos^2 \vartheta$$

The solution to this is easily seen to be $\vartheta = \tan^{-1}((a/b)^{1/3})$. You can see how ugly the solution in terms of x is going to be if in order to obtain it, you have to solve that ghastly equation. And you're not done yet, because then you have to substitute x into $(x^2 + a^2)^{1/2}\,(1 + b/x)$ in order to obtain the length of the pipe. Enjoy!

But the sec and csc are easily computed if the tangent is known simply by the usual method of drawing a right triangle in which the adjacent side is 1 and the opposite side is the known value of the tangent.

And this problem points out another advantage of using trig variables. Trigspeak, the language of trig variables, is much richer because you have six functions and their inverse functions available to express quantities, and you have to admit $\vartheta = \tan^{-1}((a/b)^{1/3})$ is SO much more elegant than whatever the expression for x turns out to be.

Two You Don't Often See

For some reason, you don't often see the following two optimization problems. One is pretty obvious — maximize the area of a right triangle with given perimeter P.

Let's see what happens if we use distance variables. We'll let B and H be the two sides, and θ be the angle between the hypotenuse and the base. The perimeter constraint is $B + H + (B^2 + H^2)^{1/2} = P$, and we want to maximize $BH/2$. Rewriting the constraint as $(B^2 + H^2)^{1/2} = P - (B + H)$ and squaring both sides, we see that $B^2 + H^2 = P^2 - 2P(B + H) + B^2 + 2BH + H^2$. Things could be worse. Simplifying, we obtain $P^2 - 2PB = H(2P - 2B)$. This can be solved for H and plugged into $BH/2$; we see that we have to maximize the function $B(P^2 - 2B)/4(P - B)$. Not a fate worse than death, the numerator of the derivative will be

$$4(P - B)(P^2 - 4B) - (-4)B(P^2 - 2B) = 8B^2 - 16PB + 4P^3,$$

which we'd have to set equal to 0. A little on the ugly side.

Let's check out how trig variables do. We'll use the length of the hypotenuse D (I have a reason for using D which you'll soon see) and the angle θ as variables. The constraint is $P = D + D\cos\vartheta + D\sin\vartheta = D(1 + \sin\vartheta + \cos\vartheta)$, and the area to be maximized is ½ $(D\cos\vartheta)(D\sin\vartheta) = \frac{1}{2} D^2 \sin\vartheta\cos\vartheta$. Solving the perimeter constraint for D and plugging it into the formula for area, we see that we'd have to maximize the function $\frac{1}{2}\sin\vartheta\cos\vartheta\,(P/(1 + \sin\vartheta + \cos\vartheta))^2$. Factoring out constants, the numerator of the derivative, which we will need to set equal to 0, will be

$$(1 + \sin\vartheta + \cos\vartheta)^2(\cos^2\theta - \sin^2\theta) - \sin\vartheta\cos\vartheta$$
$$(2\,(1 + \sin\vartheta + \cos\vartheta)(\cos\vartheta - \sin\vartheta)$$

Good news! When we set this equal to 0, one thing jumps out at you — $\cos\vartheta - \sin\vartheta$ is a factor of both terms, and the solution is the 45–45–90 triangle. Why are we not surprised! Anyway, this would undoubtedly be the case with the equation involving B above, but when you get it, until you do some algebra you won't recognize that you're looking at a 45–45–90 triangle, because when the perimeter P is given, the short side is $(\sqrt{2}/(2(1+\sqrt{2})))P$ and when you got that answer for B, you probably wouldn't know you were looking at a 45–45–90 triangle.

A definite edge to trig variables here, but catch this next act. We're going to change the problem slightly. Again, we're given a fixed amount of material P, but we're going to use it to construct both the perimeter and one diagonal of a rectangle. Our goal, of course, is to maximize the area of the rectangle.

If we go back to the equation relating the base B and height H of the rectangle, the perimeter constraint is $2(B + H) + (B^2 + H^2)^{1/2} = P$, and we want to maximize BH. Rewriting the constraint as $(B^2 + H^2)^{1/2} = P - 2(B + H)$ and squaring both sides, we see that $B^2 + H^2 = P^2 - 4P(B + H) + 4(B^2 + 2BH + H^2)$. Now things are nasty, because we don't have the $B^2 + H^2$ terms canceling on both sides. We now have to solve a really ugly quadratic to get B in terms of H — THEN multiply that by B, and differentiate it. Clearly it's time to play the useful "author's privilege" card again.

Now let's try trig variables. We'll use the same angle and the same distance, but now the reason for D appears — it's the length of one of the diagonals of the rectangle. The perimeter constraint is now $P = D + 2D\cos\vartheta + 2D\sin\vartheta = D(1 + 2(\sin\vartheta + \cos\vartheta))$, and the area to be maximized is $(D\cos\vartheta)(D\sin\vartheta) = D^2\sin\vartheta\cos\vartheta$. Solving the perimeter constraint for D and plugging it into the formula for area, we see that we'd have to maximize the function $\sin\vartheta\cos\vartheta\,(P/(1 + 2(\sin\vartheta + \cos\theta)))^2$. Factoring out constants, the numerator of the derivative, which we will need to set equal to 0, will be

$$(1 + 2(\sin\theta + \cos\theta))^2(\cos^2\theta - \sin^2\theta) -$$
$$\sin\theta\cos\theta\,(2(1 + \sin\theta + \cos\theta)(2(\cos\theta - \sin\theta))$$

And again $\cos\vartheta - \sin\vartheta$ is a factor of both terms — and the rectangle is a square. We needn't bother with the perimeter plus two diagonals problem, because that reduces to the right triangle problem we've already solved, as the rectangle with two diagonals configuration is simply the union of two congruent right triangles.

Although this is a chapter on calculus, there's a really cute proof that the right triangle with given perimeter that has maximum area is the 45–45–90 right triangle. You need to know that the geometric mean of two numbers p and q, $\sqrt{pq}$, is always less than or equal to the arithmetic mean, $(p + q)/2$ of the same two numbers. This is easy to see, just square both means. When you square the geometric mean, you get pq, and when you square the arithmetic mean, you get $(p^2 + q^2)/4 + pq/2$. Multiply both quantities by 4, and you're looking at $4pq$ and $p^2 + q^2 + 2pq$. Subtract $4pq$ from both, and you're comparing 0 and $p^2 + q^2 - 2pq = (p - q)^2$, which is always greater than or equal to zero. Since $(p - q)^2 = 0$ only when $p = q$, we see that the arithmetic mean is always greater than the geometric mean, and the two are equal only when the two numbers p and q are equal.

Anyway, back to maximizing the area of the right triangle with given perimeter. Suppose we think of the right triangle as having sides a and b and hypotenuse c. Then, as usual, $a^2 + b^2 = c^2$. Looking at the arithmetic and geometric means of the two numbers a^2 and b^2, we see that the geometric mean is ab, which is less than or equal to the arithmetic mean of $(a^2 + b^2)/2$, and the two are equal only when $a^2 = b^2$. But the area of the right triangle is $ab/2 \leq (a^2 + b^2)/4 = c^2/4$. So the maximum possible area of the right triangle is $c^2/4$, and this happens only when $a = b$, and in this case $c = \sqrt{2}a = \sqrt{2}b$, and we're looking at a 45–45–90 triangle.

Even those of us who love calculus — and I'm out there leading the parade — love it when we can do a problem that's ostensibly calculus without invoking calculus.

Four Problems, One Diagram

I love it when you can get a lot of bang for your buck. I started out trying to find the area of the smallest triangle that could be circumscribed about a given circle, and drew the following picture.

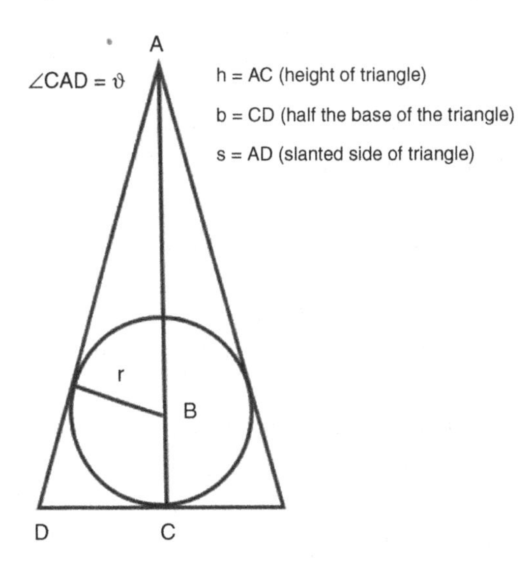

The radius of a circle is always perpendicular to the tangent at the point of contact, so $AB = h - r = r \csc \vartheta$. Therefore $h = r(1 + \csc \vartheta)$. Also $b/h = \tan \vartheta$, so $b = h \tan \vartheta = r(1 + \csc \vartheta) \tan \vartheta$, and the area of the triangle is $\frac{1}{2} h\,(2b) = r^2\,(1 + \csc \vartheta)^2 \tan \vartheta$. Differentiating and setting this equal to 0 yields the equation

$$r^2[2(1 + \csc \vartheta)\,(-\csc \vartheta \cot \vartheta) \tan \vartheta + (1 + \csc \vartheta)^2 \sec^2 \vartheta] = 0$$

This simplifies to $-2 \csc \vartheta + (1 + \csc \vartheta) \sec^2 \vartheta = 0$. Multiply by $\sin \vartheta \cos^2 \vartheta$ to obtain $-2 \cos^2 \vartheta + \sin \vartheta + 1 = 0$, and substituting $\cos^2 \vartheta = 1 - \sin^2 \vartheta$ yields $2 \sin^2 \vartheta + \sin \vartheta - 1 = 0 = (2 \sin \vartheta - 1)(\sin \vartheta + 1)$. So $\sin \vartheta = \frac{1}{2}$, and $\vartheta = 30°$. It's probably not too surprising that this the solution turns out to be an equilateral triangle.

Now suppose we look at what I consider to be the dual problem — fixed area, minimize perimeter. Look at the two equations again. Since $b/s = \sin\vartheta$, $s = b \csc\vartheta$

$$\text{area} = r^2 (1 + \csc\vartheta)^2 \tan\vartheta$$

$$\text{perimeter} = 2(b + s) = 2b(1 + \csc\vartheta) = 2r(1 + \csc\vartheta)^2 \tan\vartheta = 2x\ \text{area}/r$$

But, except for a constant factor, we already know that this is the function whose minimum occurs at $\vartheta = 30°$. I'm guessing that had we written out the value of the perimeter when we were doing the initial problem — and if you recall, we never explicitly did — we'd have seen this immediately, or close to it. But it would take someone with much greater facility in geometry than I have to recognize the fact that no matter how you circumscribe the triangle, the ratio of area to perimeter is always the same.

Now let's take it up a notch by moving into three dimensions. We'll start with a sphere and circumscribe a cone around the sphere — so the sphere is tangent to the cone at the center of the base of the cone, and there's a plane parallel to the base of the cone where the sphere is tangent to every point on the surface of the cone that belongs to the plane. Just imagine an ice cream cone in which the spherical scoop of ice cream was so small that it slipped down into the cone, there will be a circle of ice cream on the cone — that's the place where the ice cream sphere is tangent to the ice cream cone.

The diagram above works perfectly well for this situation — it's a cross-section perpendicular to the base of the cone going through the vertex of the cone (the point A in the diagram). First question — what's the smallest volume of such a cone?

For this problem, using the notation in the diagram, b is the radius of the cone, h is the height of the cone, and s is the slant height of the cone. The cone volume is $\pi\, b^2h/3$. However, we've worked out all these quantities before, so the cone volume is $(\pi r^3/3)\ (1 + \csc\vartheta)^3 \tan^2\vartheta$. Differentiating this with respect to ϑ and setting the result equal to 0 yields

$$0 = 3\ (1 + \csc\vartheta)^2 \tan^2\vartheta\ (-\csc\vartheta \cot\vartheta) + 2\ (1 + \csc\vartheta)^3 \tan\vartheta \sec^2\vartheta$$

Divide out by the factor $(1 + \csc\vartheta)^2 \tan\vartheta$, and observe that $\tan\vartheta \cot\vartheta = 1$. This leaves $0 = -3 \csc\vartheta + 2 (1 + \csc\vartheta) \sec^2\vartheta$. Multiply by $\sin\vartheta \cos^2\vartheta$ to obtain $0 = -3 \cos^2\vartheta + 2 (\sin\vartheta + 1)$. Now substitute $\cos^2\vartheta = 1 - \sin^2\vartheta$, getting $0 = 3 \sin^2\vartheta + 2 \sin\vartheta - 1 = (3 \sin\vartheta - 1)(\sin\vartheta + 1)$, and we see that the cone with the smallest volume is characterized by $\sin\vartheta = 1/3$.

And once again, let's look at the dual problem, where we want to find the cone with the smallest surface area — that's the lateral area (the sides of the cone) plus the base. Using the labels in the diagram, the lateral area is πbs and the base is πb^2. So the total surface area is $\pi b (s + b)$. We've already observed that $s + b = b(1 + \csc\vartheta)$, and so the total surface area is $\pi b^2 (1 + \csc\vartheta)$.

Surprise! This is just $\pi r^2 (1 + \csc\vartheta)^3 \tan^2\vartheta$, which is simply a constant multiple of the volume, so once again we needn't bother with the differentiation. The same cone that minimized volume also minimizes total surface area.

Trig is SO much more than just using it for the classic "find the height of a tree" problems. Back when I was taking trig, the book had a table of the values of all the trigonometric functions for every angle that was measured in a whole number of degrees, from 1° through 89°. These numbers were given to four decimal places. After working my way through trig, in which you learn a number of tricks for finding the values of the trigonometric functions for angles such as 15° (and a few others by using the half-angle and addition and subtraction formulas), I wondered — how do they find sin 19° accurate to four decimal places? Do they hire people to construct really large right triangles with a 19° angle and measure them really accurately?

We'll see the answer to that in Chapter 8. But I wonder nowadays if students, who have calculators that give these numbers to eight or more decimal places — and do so even if you ask it to find sin 19.3784°, ask similar questions. I sure hope so.

Bibliography

[1] C. Snow, *The Two Cultures*. London, UK: Cambridge University Press, 1959.

Chapter 7

Seduced by Complex Numbers

Rene Descartes' Imaginary Contribution

Complex numbers, as we use them today, can trace their lineage more than 2,000 years, back to Hero of Alexandria. However, they started gaining more currency in the Middle Ages, as the Italian algebraists of that era raced to discover solutions to the general cubic and quartic polynomials. But even though the computational rules for complex numbers had been established by the start of the 17th century, Rene Descartes was unconvinced of their worth. Descartes was not just responsible for analytic geometry, he was responsible for the term "imaginary number" for the square roots of negative numbers. He chose the term as a derogatory one because he thought that imaginary numbers were mathematical fictions that had no use in the real world.

And, boy, was he wrong!

But let's dispense with this concept of "mathematical fiction." I'm not sure that any number is anything other than a mathematical fiction — because I can't define it. I used to teach courses in Mathematics for Elementary School Teachers, and I would tell them that I had no idea what "three" was. All I could say was that "three" was the common ground between a set consisting of three children and a set consisting of three cookies. I can tell you a lot of properties of "three" — how it behaves when combined with other numbers via arithmetical operations, but I don't think I can tell you what "three" is.

And that's pretty much true of any really fundamental mathematical concept. But authors of science fiction (and other areas as well) talk about "the willing suspension of disbelief" when it comes to accepting the premise of a plot, and it's sort of the same with mathematics. It's really hard — maybe even impossible — to define the really basic mathematical concepts. Euclid may have been the first mathematician to realize this difficulty, he defines a point to be "that which has no part." Early in the 20th century, Bertrand Russell and Alfred North Whitehead tried to come to grips with this, and created a work called *Principia Mathematica* [1], a riff on Isaac Newton's *Philosophiae Naturalis Principia Mathematica* [2]. I didn't really care about the foundations of mathematics, I just wanted to work on problems I found interesting, and so I took somebody else's word that Russell and Whitehead didn't get to $1 + 1 = 2$ until about page 800.

The Work of Man

Leopold Kronecker was a 19th-century German mathematician who famously declared, "The integers are the work of God, all else is the work of Man" [3]. So how do we get to a point where we can actually talk about something called "the square root of minus 1"?

Yes, they knew the rules for computing with expressions of the form $a + bi$ in the 16th century — but how do we know we're actually talking about something? Mathematics had faced this same dilemma before — first with zero, and then with negative numbers — but you can put real-world interpretations on both of them. In fact, money is a really good model for these concepts, as zero corresponds to having no money, and negative numbers correspond to being in debt.

And that's probably akin to what Descartes had in mind. But there's a more serious issue than just "mathematical fiction" associated with the square root of -1. What if it's more than "mathematical fiction" — it's total fiction in the sense that no such mathematical entity exists?

One of my former professors encountered this particular stumbling block. Without going into gory details, we can think of the real numbers as a number system of dimension one. It turns out that the complex numbers are a similar system of dimension two. In the 19th century, mathematicians came up with quaternions, which were a similar system of dimension four, and Cayley numbers, which were a system of dimension eight.

As you can probably guess, the next number system up the line would be of dimension sixteen, and my professor spent some time working out theorems that would be true of such a system. Just before he was ready to submit his paper, another mathematician managed to demonstrate that there were no such systems of dimension sixteen — or higher.

Here's how mathematicians got around that problem for the complex numbers. They defined a system of ordered pairs (a, b) of real numbers which obeyed the following rules of addition and multiplication.

$(a, b) + (c, d) = (a + c, b + d)$ example: $(2, 3) + (1, -7) = (3, -4)$
$(a, b) \times (c, d) = (ac - bd, ad + bc)$ example: $(2, 3) \times (1, -7) = (23, -11)$

They managed to show this system had all the good stuff concerning number systems — the commutative, associative, and distributive laws, the presence of an additive identity ($(a, b) + (0,0) = (a, b)$), the presence of a multiplicative identity ($(a, b) \times (1, 0) = (a, b)$), and the existence of additive and multiplicative inverses.

You can see that the real numbers appear in this system as all ordered pairs of the form $(a, 0)$, and that the product $(0, 1) \times (0, 1) = (-1, 0)$. And it's not hard to realize what motivated this system — the complex number $a + bi$ corresponds to the ordered pair (a, b).

This may strike you as a distinction without a difference — but to a mathematician, it's significant. Talking about the square root of -1 is NOT the same as exhibiting a mathematical object in which it makes sense. My former professor who proved theorems about number systems of dimension sixteen might ruefully concur.

So here's what it might have looked like to a mathematician of the Middle Ages. If you have a linear equation — a polynomial of degree one — with real coefficients, it has real solutions. A polynomial of degree two, however, might have these weird things as solutions. Could weirder things be out there as solutions of polynomials of degree three or higher?

The solution to the general cubic equation was known by 1510, and the solution to the general quartic by 1540. So far, no new weird things were out there, as the solution to both of the above could be expressed in terms of real and complex numbers. But here progress stalled, because as Galois was to discover some two and a half centuries later, there is no general solution to the quintic (a polynomial of degree five) in terms of powers and roots.

However, almost at the same time Carl Friedrich Gauss proved the Fundamental Theorem of Algebra — every polynomial of degree n with real coefficients has n real and complex roots. There are no more weird things out there — at least as far as solutions to polynomials are concerned.

But how did you find those solutions?

The Complex Plane

Things were coming together. Within a decade, the Swiss mathematician Jean Argand had combined Descartes' idea of a coordinatized plane with the complex numbers to create a graphical representation in which the point (x, y) in the coordinate plane represented the number $x + iy$. The distance of that point from the origin (0,0) was defined to be $|x + iy|$ — the absolute value $x + iy$.

Polar coordinates were invented in the 17th century, and complex numbers were known before then. So it's somewhat surprising that it took until early in the 19th century for the French mathematician Jean Argand to meld the two, given that it seems so natural to depict $x + iy$ as the point (x,y) in a two-dimensional plane. Of course we have the advantage of hindsight, and we typically write complex numbers as $z = x + iy$. But maybe they used different letters; certainly the use of x and y should serve as a trigger for inspiration.

Isaac Newton is known for a LOT of things, but as far as I know, no great insights into the complex numbers. He had, however, come up with the idea of polar coordinates — which look a lot like Argand's plane. If one thinks of r in polar coordinates as $|x + iy|$, and ϑ as the angle between the positive x-axis and the line from (0,0) to (x,y) — the same in both polar coordinates and the complex plane.

Once you decide to depict a complex number as a point in the complex plane, and with the idea of polar coordinates in mind, a lot of things naturally fall into place. When we write (x,y) in polar coordinates (r,θ), $r = (x^2 + y^2)^{1/2}$. But if we think of (x,y) as representing the point $x + iy$, then r is immediately seen to be $|x + iy|$. And since $x = r \cos\theta$ and $y = r \sin\theta$, we immediately have the representation $x + iy = |x + iy|(\cos\theta + \mathrm{i} \sin\theta)$.

Now we see an interplay between the geometry of polar coordinates and the algebra of complex numbers. If we look at the sum $(x + iy) + (u + iv) = (x + u) + i(y + v)$, we recognize the parallel between addition of complex numbers and addition of vectors in the plane — and so we can add complex numbers using the parallelogram law for addition of vectors. But the really interesting situation comes when we look at the geometric representation for multiplication. If we use the polar representation for two complex numbers, if $z_1 = r_1(\cos\theta_1 + i\sin\theta_1)$ and $z_2 = r_2(\cos\theta_2 + i\sin\theta_2)$, then $z_1z_2 = r_1r_2(\cos(\theta_1 + \theta_2) + i\sin(\theta_1 + \theta_2))$, by DeMoivre's Theorem. This means that once we have located z_1 in the complex plane, we can multiply it by z_2 by stretching the distance from the origin by the factor r_2, and rotating the result of the stretching counterclockwise by an angle θ_2.

In particular, multiplying by a complex number whose absolute value is 1 is simply a rotation. This leads to a beautiful result that is part algebra and part geometry.

Euler's Formula

The formula $e^{ix} = \cos x + i\sin x$ is known as Euler's Formula. I'm not sure what the usual proof is, but here's a quick one if you're willing to accept that, from the standpoint of differentiation, complex constants behave like real constants. I don't think this comes under the heading of "willing suspension of disbelief" — it's not like I'm asking you to accept superheroes or time travel.

Let $f(x) = \cos x + i \sin x$. Then $f'(x) = -\sin x + i \cos x = i(\cos x + i \sin x)$. Therefore $f(x)$ satisfies the differential equation $f'(x) = i f(x)$, and (here's where I'm using my assumption) the solution to this is known to be $f(x) = Ce^{ix}$, where C is a constant. But $f(0) = \cos 0 + i \sin 0 = 1 = Ce^{i0} = C$, and so $f(x) = e^{ix}$.

Did I mention that one of the core attractions of radian measure is that it enables derivatives of trig functions to be clean? We know that the derivative of $\sin x$ is $\cos x$ — but that's because we measure the angles in radians. This allows us to prove that $\lim_{x \to 0} \frac{\sin x}{x} = 1$, the key limit in establishing the fact that the derivative of $\sin x$ is $\cos x$.

So Euler's formula uses radian measure, and when we substitute $x = \pi$, we see that $e^{i\pi} = -1$. Certainly an intriguing result, but when you add 1 to both sides, you see that $e^{i\pi} + 1 = 0$. This is almost universally acknowledged, among mathematicians at any rate, to be the most beautiful formula in mathematics. Think of it, collected in one formula are 0 (the additive identity), 1 (the multiplicative identity), e (the number that is the base of the formulas for natural growth), π (the core number of geometry), and i (the basic imaginary number). There was a day in December 1956 when the Sun Records studio in Memphis, Tennessee saw four titans of rock'n roll all together — Elvis, Johnny Cash, Jerry Lee Lewis, and Carl Perkins [4] — but who will remember them 100 years from now? As long as mathematics is being done anywhere by anyone — or anything – 0, 1, e, π and i will occupy the summit of whatever mountain range contains the most important constants.

But that's not the formula that I find most intriguing, at least in the offspring of Euler's Formula. The formula that most enchants me comes when we substitute $\pi/2$ for x, and do a little extra work. Euler's Formula shows that $e^{i\pi/2} = \cos \pi/2 + i \sin \pi/2 = i$. Now raise both sides to the power i. The right side becomes i^i, and the left side becomes $(e^{i\pi/2})^I = e^{-\pi/2}$, which is approximately 0.20788. And how bizarre is that? The most fundamental imaginary number — raised to the power of the most fundamental imaginary number, turns out to be a little more than one-fifth. I don't know whether that's amazing or disappointing — but it certainly is unexpected. What would you have suspected i^i was going to be?

And while we're on the subject, what about the ith root of i? Well, following along with the meaning of roots with real numbers, the ith root of will be $i^{1/i} = i^{-i} = 1/i^i = e^{\pi/2}$ — approximately 4.81048. Hard to believe.

De Moivre's Theorem

Frankly, I'm not even sure why this merits the title of a theorem. It says that if n is an integer, then $(e^{ix})^n = \cos nx + i \sin nx$. Usually this is proved by induction, starting from Euler's formula (the case $n = 1$). But why bother? If you accept that the rules of exponentiation apply to complex numbers as well (and why wouldn't you?), then $(e^{ix})^n = e^{i(nx)} = \cos nx + i \sin nx$ by Euler's Formula. In fact, this argument seems a whole lot more powerful to me than De Moivre's Theorem, because it applies for ANY value of n, real or complex, not just integers.

And while we're on the subject, look at how easy it is to prove the addition formulas for sine and cosine in trigonometry. The proof that I saw used a diagram something like this.

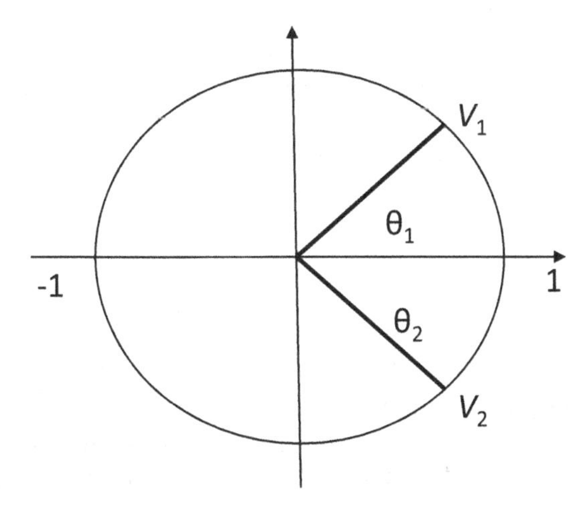

I'm not going to reprise the proof here, just give you the general idea. You compute the distance between the points V_1 and V_2 using the distance formula, $V_1 = (\cos \vartheta_1, \sin \vartheta_1)$ and $V_2 = (\cos \vartheta_2, \sin \vartheta_2)$. You then rotate the entire circle by an angle ϑ_2 counter-clockwise, the point V_2 now occupies the position $((\cos(\vartheta_1 + \vartheta_2), \sin(\vartheta_1 + \vartheta_2))$, while the point V_1 is now sitting on (1,0). The rotation doesn't change the distance between V_1 and V_2, so you equate the two distances, do some algebra, and emerge with the addition formula for the cosine.

But your work is not yet done, In order to get the addition formula for the sine, the standard proof now requires that you use the addition formula you've just developed for the cosine and the fact that sin $(\vartheta_1 + \vartheta_2) = \cos(\pi/2 - (\vartheta_1 + \vartheta_2))$ and a bunch more algebra and trig. This is work — a LOT of work.

But look how easy it becomes using Euler's Formula!

$$e^{i(\theta_1+\theta_2)} = \cos(\theta_1 + \theta_2) + i\sin(\theta_1 + \theta_2)$$

$$= e^{i\theta_1} e^{i\theta_2} = (\cos\theta_1 + i\sin\theta_1)\,(\cos\theta_2 + i\sin\theta_2)$$

$$= (\cos\theta_1 \cos\theta_2 - \sin\theta_1 \sin\theta_2) + i\,(\cos\theta_1 \sin\theta_2 + \sin\theta_1 \cos\theta_2)$$

Equate real and imaginary parts, and we're done.

Let's spend a little time on the function $f(x) = e^{ix}$, where x is real. We know that the real exponential function e^x is one-to-one — if x and y are different real numbers, e^x and e^y are different real numbers as well. But that's not true for the function e^{ix}; it's periodic with period 2π because both $\cos x$ and $\sin x$ are periodic with period 2π. This surprising result, combined with De Moivre's Theorem, enables us to find all the nth roots of 1.

The *n*th Roots of 1

Let's suppose that some time in the middle of the 17th century, you have been tasked with the job of finding the nth roots of 1, where n is an integer. You suspect — although considering the stage to which algebra has been developed, you cannot be sure, that there are n of them. You know that there are two square roots of 1, i and $-i$ — and these numbers are the solutions of the equation $x^2 - 1 = 0$.

You surmise that the cube roots of 1 are the solutions of the equation $x^3 - 1 = 0$. Well, there is a formula that enables you to factor the difference of two cubes, and you obtain $x^3 - 1 = (x - 1)(x^2 + x + 1)$. The Quadratic Formula has been known for some time, and you find the solutions to $x^2 + x + 1 = 0$ are $(-1 \pm \sqrt{3}i)/2$. A little work enables you to verify that these are indeed cube roots of 1; together with 1 you've found 3 cube roots of 1.

The fourth roots of 1 are a slam dunk — or whatever the 17th-century equivalent of a slam dunk is. You can factor $x^4 - 1 = (x - 1)(x + 1)(x^2 + 1)$, and the four roots of these are 1, –1, and –*i*. You now pause for reflection, and realize that you can probably find the five fifth roots of I, as you can factor $x^5 - 1$ as the product of $x - 1$ and a fourth-degree polynomial — for which there is a general solution. You can also factor $x^6 - 1 = (x^3 - 1)(x^3+1)$ — and you can find the roots for both of these cubics. But you're going to be stuck at $x^7 - 1$, which is the product of $x - 1$ and a sixth-degree polynomial for which you have no tools with which to work.

But look at how this problem becomes a genuine slam dunk after the introduction of the complex plane! If a complex number z is written as $z = r(\cos\theta + i\sin\theta)$, De Moivre's Theorem shows that the complex number $r^{1/n}(\cos(\theta/n) + i\sin(\theta/n))$ is an nth root. But because the same complex number admits any representation of the form $r(\cos(\theta + 2\pi k) + i\sin(\theta + 2\pi k))$, where k is an integer, we immediately see that any expression $\mathrm{r}^{1/n}(\cos((\theta + 2\pi k)/n) + I\sin((\theta + 2\pi k)/n))$ is also an nth root for any integer k, and if we let $k = 0, 1, 2, \ldots, n - 1$ we get the nth roots of the complex number z. The geometry of the nth roots of 1 is particularly attractive, as they are all equally spaced on the circumference of the unit circle, and the point $1 + 0i$ is always one of the roots.

Partial Fractions

Math has its tedious moments, and high on the list for most students would be the topic of partial fractions — the process of decomposing a rational function into a sum of simpler ones. The first reason that it's tedious is that it's often introduced in Precalculus, where no immediate application is given other than to say, "Watch out for this in Calc II." This is somewhat akin to telling a 7-year-old to eat her vegetables to prevent heart trouble when she is in her seventies.

For those who have never seen it, or have mercifully forgotten it, here's a simple example to show the two basic techniques used in this process.

Suppose we want to find constants A and B which make the following expression an identity.

$$\frac{2x-11}{2x^2-7x+5}=\frac{A}{x-1}+\frac{B}{2x-5}$$

Multiply both sides by $(x-1)(2x-5)$, noticing that this is the factorization of the denominator of the left-hand side. The result is the equation $2x - 11 = A(2x-5) + B(x-1)$. The first technique for finding the constants A and B is known as the knock-out process. Since this equation is true for all values of x, choose a value of x which "knocks out" one of the terms. In particular, if we let $x = 1$, it knocks out the second term (since $x-1 = 0$), leaving us with the equation $-3A = -9$, and we quickly see $A = 3$. Similarly, letting $x = 5/2$ knocks out the first term, and we're left with the equation $3B/2 = -6$, so $B = -4$. Plugging these values in for A and B and computing the sum will verify that this decomposition is correct.

The other method, seeing as the equation $2x-11 = A(2x-5) + B(x-1)$ is true for all x, is to equate coefficients of each term of the two polynomials. For the coefficient of x, we get the equation $2A + B = 2$, and for the constant term, we get the equation $-5A - B = -11$. Solving this, we not surprisingly find that the solution is $A = 3$ and $B = -4$.

In this situation, the knock-out method is preferable, because it's much easier and less prone to error to solve two different equations, each in one unknown, that a single set of two equations in two unknowns. When one has more unknowns, the student is advised to first see if there's a knock-out available, and then use the value or values for those unknowns, equate coefficients, and obtain a system of equations in the other unknowns.

So let's look at a more complicated problem.

Suppose we want to find constants A, B, C, and D to make the following an identity.

$$\frac{6x^3+10x^2+26x+7}{\left(x^2+2x+5\right)\left(x^2+4\right)}=\frac{Ax+B}{x^2+2x+5}+\frac{Cx+D}{x^2+4}$$

Multiply both sides by the left-hand denominator to obtain the equation

$$6x^3+10x^2+26x+7=(Ax+B)\left(x^2+4\right)+(Cx+D)\left(x^2+2x+5\right)$$

Not seeing any useful real numbers to substitute, you resort to equating coefficients to obtain the following system of four equations in four unknowns.

$$A+C=6$$

$$B+2C+D=10$$

$$4A+5C+2D=26$$

$$4B+5D=7$$

Not an attractive prospect — but look what happens if we think a little outside the box and try knocking out a term using complex numbers. If we let $x = 2i$, $x^2 + 4 = 0$. Substituting this into the left side of the polynomial equation yields $-48i - 40 + 52i + 7 = -33 + 4i$. Substituting this into the right side of that equation gives $(2iC + D)(-4 + 4i + 5) = (2iC + D)((1 + 4i) = (2C + 4D)i + (D - 8C)$. Equating real and imaginary parts, we now are faced with two equations in two unknowns

$$2C+4D=4$$

$$D-8C=-33$$

The solution is easily seen to be $C = 4$, $D = -1$. Substituting these values into the first and fourth equations above quickly results in $A = 2$, $B = 3$. Sweet!

Whenever I teach differential equations, I show students this trick — most of them haven't seen it and I don't want them spending their time solving a four by four system of linear equations. It's also very useful for Laplace Transforms.

And speaking of differential equations....

Complex Numbers and Differential Equations

Even though you may be unfamiliar with differential equations, if you have nothing but a single semester of calculus you should be able to wade your way through this section. Let's try to solve the differential equation $y'' + 4y' + 13y = 0$. We have a stroke of inspiration and decide to try for a solution of the form $y = e^{rx}$, where r is a constant. Since $y' = re^{rx}$ and $y'' = r^2e^{rx}$, on substituting these expressions and factoring, we see that $e^{rx}(r^2 + 4r + 13) = 0$. Since e^{rx} is never equal to 0, we can divide both sides of the equation by it, obtaining the quadratic equation $r^2 + 4r + 13 = 0$.

The Quadratic Formula (or completing the square, which is easier here) shows that the roots of this equation are $-2 \pm 3i$, which means that two solutions of this differential equation should be $e^{(-2+3i)x}$ and $e^{(-2-31)x}$. But what are we to make of this, as neither are real-valued functions? And, if they tried this approach in the era B.A. (Before Argand), what did they make of it?

Granted, when you have the all-star lineup of seventeenth and 18th century mathematicians looking at this, it's clear they would have somehow worked out that $e^{-2x} \cos 3x$ and $e^{-2x} \sin 3x$ were solutions to the differential equation. But they might have been a little perplexed as to how to justify this, other than the obvious "it works, so why worry?"

However, we know that

$$\begin{aligned} e^{(-2+3i)x} &= e^{-2x}(\cos 3x + i \sin 3x) \quad \text{and} \\ e^{(-2-3i)x} &= e^{-2x}(\cos(-3x) + i \sin(-3x)) \\ &= \overline{e^{(-2+3i)x}} = e^{-2x}(\cos 3x - i \sin 3x). \end{aligned}$$

Let's call the first solution y_1 and the second solution y_2. We know that constant multiples of solutions are solutions, and sums of solutions are solutions, so $e^{-2x}\cos 3x = (y_1 + y_2)/2$ is a solution, as is $e^{-2x}\sin 3x = (y_1 + y_2)/2i$.

If you are a theorist, you might be a little nervous about the way we've happily assumed that some of the stuff we've done, such as differentiating complex-valued functions, works the same way that it does for differentiating real-valued functions. Not to worry, theorists have managed to justify this long before you, or your great-grandparents, were born. It may seem that I'm denigrating theory here, but that's not the case — there are some instances where mathematics that seemed intuitively reasonable went astray — and theory was needed to escape from the predicament.

A Cutie from Multivariable Calculus

I've always maintained that calculus should be taught initially by emphasizing what it can do rather than what its limitations are — because sooner or later we will encounter those limitations in the natural course of the development of calculus. And sooner or later, the calculus student will be faced with a function such as e^{-x^2} which has no simple antiderivative. But it is fascinating to learn that there are techniques which enable a number of definite integrals to be computed **exactly** without being able to find an antiderivative and apply the Fundamental Theorem of Calculus.

I bolded the word "exactly" in the previous sentence because there are numerical methods of greater or lesser simplicity enabling the evaluation of any integral to any desired degree of accuracy. I have nothing against numerical analysis. Indeed, an approximate answer is far better than no answer, and knowing an approximate answer and how accurate the approximation is in many cases the best we can do — and fortunately, it's good enough for practical purposes.

Seductive, however, is what numerical analysis is not — at least for me. It's meat and potatoes math.

I mentioned e^{-x^2} earlier because it shows up in the first knock-your-socks-off definite integral most students see, usually in multivariable calculus because you need to know how to transform a double integral from rectangular to polar coordinates. I don't know who's responsible for discovering this particular trick, but it's a honey.

Let's suppose we want to compute $C = \int_0^\infty e^{-x^2} dx$. This particular integral is extremely important in both probability and statistics, perhaps the trick of evaluating it was discovered in conjunction with the development of these subjects. But, because x is known in the trade as a dummy variable, it's also true that $C = \int_0^\infty e^{-y^2} dy$. Now watch the following magic trick.

$$\begin{aligned} C^2 &= \int_0^\infty e^{-x^2} dx \int_0^\infty e^{-y^2} dy \quad \text{this is straightforward} \\ &= \int_0^\infty \int_0^\infty e^{-x^2} e^{-y^2} dy\, dx \quad \text{because the } y \text{ integral is constant w. r. t. } x \\ &= \int_0^\infty \int_0^\infty e^{-(x^2+y^2)} dy\, dx \end{aligned}$$

Notice that in the xy-plane, the region of integration is the first quadrant. If we transform this integral to polar co-ordinates, in which the differential element of area is $r\,dr\,d\vartheta$, the integral now becomes

$$= \int_0^{\pi/2} \int_0^\infty e^{-r^2} r\,dr\,d\theta$$

and the inner integral can be easily evaluated!

$$= \int_0^{\pi/2} \frac{1}{2} d\theta = \pi/4$$

So $C = \sqrt{\pi}/2$. I have nothing but admiration for the genius, whether sung or unsung, who first came up with this, but it fades in comparison when one considers what one can do with the Residue Theorem.

The Residue Theorem

The Residue Theorem is a little beyond the scope of this book, but not much. For me, it was the highlight of the first semester of a course in complex variables, and now students who embark upon calculus in their junior year of high school could encounter this as soon as they enter college.

The Residue Theorem would certainly make my list of the Ten Most Seductive Theorems in Mathematics (although I must confess I haven't made out the list), simply because it is so powerful and so completely and totally unexpected. It enables one to integrate some definite integrals which one could not do by finding an anti-derivative, because the integrands in the problem have no simple anti-derivative (one which can be expressed using a finite number of powers, roots, logarithmic and exponential functions, and trigonometric and inverse trigonometric functions).

Take the following integral.

$$\int_0^\infty \frac{\cos x}{x^2+1}dx = \frac{\pi}{2e}$$

Yes, I'm teasing the Residue Theorem here, but it evaluates this totally real integral by taking a trip into the complex plane. That's even more of a leap — IMHO — than what we saw in the previous integration where we took a trip into the real plane and used polar coordinates. Moreover, the trick we just used is limited in applicability, whereas the Residue Theorem has much greater scope.

My favorite application of the Residue Theorem — another tease — is that it can be used to prove one of the most attractive of Euler's many formulas.

$$\sum_{n=1}^{\infty} \frac{1}{n^2} = \frac{\pi^2}{6}$$

I must admit that I have absolutely no idea how to prove this result without using the Residue Theorem, but Euler must have, as he did it maybe a century or so before the Residue Theorem became known. So I looked it up. It's actually accessible to a second-semester calculus student — when Euler proved it, he cited an approach involving what he called quadratures of circles, but which is now called squaring the circle. This is a classic problem from ancient Greece that involves constructing a square whose area is equal to that of a given circle. It cannot be done using just Euclidean geometry, but there are other approaches, and I'm guessing Euler knew them.

There's a curious codicil to the above formula. We'll see in the chapter on infinite series that the above series is known as a p-series with $p = 2$ (if you've had second-semester calculus, you know this). There are techniques for computing the sums of other p-series, of which $p = 4$ is one. Euler nailed that one, too — the sum is $\pi^4/90$. As far as I know, no one has developed a formula for the exact sum of the p-series with $p = 3$. Every so often, I go back and look at the Residue Theorem proof for the p-series with $p = 2$, and try to tweak it to give the result for $p = 3$. So far, my bid to enter the history books via this particular back door has met with complete failure, but at least I'm not the only one.

Bibliography

[1] A. Whitehead and B. Russell, *Principia Mathematica*. Cambridge, UK: Cambridge University Press, 1910.

[2] I. Newton, *Philosophiae Naturalis Principia Mathematica*. Originally published in Latin in 1687. Online at https://www.maths.tcd.ie/pub/HistMath/People/Newton/Principia/Bk1Sect1/PrBk1St1.pdf.

[3] MacTutor. Online at https://www.maths.tcd.ie/pub/HistMath/People/Newton/Principia/Bk1Sect1/PrBk1St1.pdf.

[4] Wikipedia. Online at https://en.wikipedia.org/wiki/Million_Dollar_Quartet_(musical).

Chapter 8

Seduced by Infinite Series

In writing a book, the chapters fall into three categories — the ones you look forward to writing, the ones you just write, and the ones you dread writing (fortunately, none of those in this book). This chapter is definitely in Category 1.

If you've never seen this topic before, you may wonder why it's so important. It all goes back to a question that puzzled me when I was in high school. I've mentioned that the textbooks of my era all had tables for important functions in the back of the book, so you could actually compute reliable numerical answers for practical problems. I wondered how they computed the sine of 19 degrees to four decimal place accuracy. All I could think of was that somewhere they constructed a very large right triangle with a 19 degree angle and measured it so they could use the fact the sine of an angle in a right triangle is the ratio of the opposite side to the hypotenuse.

And then one day, I can't remember where, I saw the following formula

$$\sin x = x - x^3/3! + x^5/5! - x^7/7! + \cdots$$

I was in high school at the time, and although I knew that you could use that formula to compute the sine of 19 degrees, I couldn't imagine where that formula came from. Sines were trigonometry, the stuff on the right side of the equation was algebra, and I simply could not imagine how the two could be united. True, there were some formulas I'd seen in trig that combined both algebra and trig, such as $\cos(\tan^{-1}(x)) = 1/(x^2 + 1)^{1/2}$, but that infinite sum, with its beautiful easily-described pattern, appeared as if from another world.

Geometric Series

The first exposure many people have to infinite series is by way of Xeno's Paradox. Xeno argued as follows: if the arrow has to travel halfway to its target, then half of the remaining distance, then half of the remaining distance, and so on, how does the arrow ever manage to reach its target?

Well, somehow it does — unless it was badly aimed, but even so it gets somewhere, if not to the target. The obvious way to resolve this is to think of the total distance as 1. Xeno's first "halfway" amounts to ½ of the total distance, the second to ¼ of the total distance, the third to 1/8, and so you are looking at adding up the infinite series $½ + ¼ + 1/8 + \cdots$, which is usually condensed using sigma notation as

$$\sum_{n=1}^{\infty} \frac{1}{2^n}$$

Ok, everybody KNOWS that this sum is 1. As n gets larger and larger, the sum of the first n terms gets closer and closer to 1. We KNOW that — but that's not a proof. In fact, we don't even have a definition of what the sum of this series is if we were just looking at the series itself. Or, even better, some series with unknown terms. How can we evaluate what the sum of a series is?

Let's take another look at that phrase, "as n gets larger and larger, the sum of the first n terms gets closer and closer to 1." The key is to quantify "larger and larger" and "closer and closer." I have no idea who did this first, but it was genius. Here's the definition.

A series

$$\sum_{n=1}^{\infty} a_n$$

has sum S if for every $\varepsilon > 0$ there is an integer N such that $n \geq N$ implies $|S - (a_1 + a_2 + \cdots + a_n)| < \varepsilon$.

This definition is an offshoot of the definition of the limit of a sequence, which has probably terrorized more aspiring math students than any other — it absolutely traumatized me as an undergraduate. I saw it first my sophomore year in a class taught by the world-class mathematician Shizuo Kakutani. I somehow stumbled through the first semester, getting a 76, improved to an 80 the second semester, and formed the impression that if I were to succeed in mathematics, it would only be through studious avoidance of courses that featured epsilonics. That's the art of constructing proofs using definitions resembling this one, which features the Greek letter epsilon, similarly used, in a starring role.

Maybe there is something to the theory that if you let your subconscious mull over something, it can achieve breakthroughs. I was forced to confront that definition again my first year in graduate school. Having it sit in my subconscious for several years maybe had an effect, for on the second day of class, while walking near the Campanile (that large tower on the Berkeley campus), I experienced one of the two or three epiphanies I have had in my life. I totally got that definition.

Let's see how it works for the series that sums the powers of ½. If we let $s_n = ½ + ¼ + \cdots + (1/2)^n$, then it's easy to see that $½\, s_n = ¼ + 1/8 + \cdots + (1/2)^{n+1}$. Subtracting the second expression from the first, we have $s_n - ½\, s_n = ½\, s_n = (½ + ¼ + \cdots + (1/2)^n) - (¼ + 1/8 + \cdots + (1/2)^{n+1}) = ½ - (1/2)^{n+1}$, as a lot of terms cancel. Multiplying by 2, we see that $s_n = 1 - (1/2)^n$. We've now got the algebra out of the way, so let's try to show that the sum of the powers of ½ is 1 by using the definition.

Almost all proofs begin with the phrase "let $\varepsilon > 0$." That "tees up" the proof by setting a prescribed measure of closeness to the presumed sum (in this case, 1). We want to show that we can find an integer N such that if $n \geq N$, every sum of the first n terms (which we denoted by s_n) is closer to 1 than ε.

But this isn't hard! The algebra above shows that $1 - s_n = (1/2)^n$. So, if we choose N such that $(1/2)^N < \varepsilon$, which can be done simply by choosing N so large that $2^N > 1/\varepsilon$, then $n \geq N$ implies that $|1 - s_n| = (1/2)^n \leq (1/2)^N < \varepsilon$, and we're done.

A geometric series is the sum of the powers of r, generalizing the sum of the powers of r. It's not hard to show, using a similar argument, that as long as $0 < |r| < 1$, the sum of the geometric series is $1/(1 - r)$, and if $|r| \geq 1$, the series simply doesn't have a sum — the sum of the first n terms eventually get bigger than any number, so the series can't have a sum that's a real number.

Instead of thinking of the geometric series as consisting of powers of a number r, we could also think of it as consisting of powers of a variable x. Doing this enables us to view a geometric series as a function of x whose domain consists of all those numbers x such that $0 \leq |x| < 1$. When viewed as a function, the domain is the open interval $(-1, 1)$; this is known as the interval of convergence for the series.

Questions

There are a bunch of questions related to summing an infinite series. The first question is — which infinite series can be summed? There are two ways in which an infinite series can fail to have a sum. We've encountered the first — the partial sums (a partial sum is the sum of the first n terms of a series) get too large. The second is that they bounce around, such as with the series $1 - 1 + 1 - 1 + \cdots$. The partial sums bounce back and forth from 1 to 0 to 1 to 0 and so on. Assuming that this series has a sum we call S leads to the following result, which you'll sometimes encounter as "evidence" that mathematics doesn't make any sense.

$$S = 1 - 1 + 1 - 1 \cdots = 1 - (1 - 1 + 1 \cdots) = 1 - S$$

Solving $S = 1 - S$ gives the answer $S = ½$.

The second question is — how many terms must you take in the infinite summation to get close to the actual sum? After all, the geometric series with first term r and ratio r has sum $r/(1 - r)$ as long as $|r| < 1$ – but we're NEVER going to use the series to compute the value of $r/(1 - r)$, which we can do by long division if necessary. We are, however, going to use the series for sin 19° to compute sin 19°, so it would be nice to know how many we need.

This is a really important question, and is handled by something called the error term. The error is the difference between a partial sum and the actual value of the series. It's a measure of how close the partial sum, which you can compute using the basic arithmetic operations (adding, subtracting, etc.) is to the actual sum of the series. The great results in this area are the ability to estimate the largest the error could possibly be. If you know the size of the error is less than 0.005, for example, you know that your partial sum is accurate to the hundredths place. We'll see several of these.

Finally, a related question is — we measure speed of convergence to be how many terms we need in the partial sum to get within a desired maximum error. So how do we find series that converge more rapidly than others? There are a number of great results in this area, mostly beyond the scope of this book.

Tool #1 — The Thin Sub-sequence Test

Throughout this section, we'll be looking at the series

$$\sum_{n=1}^{\infty} a_n$$

There are two really nice results on the convergence of series that I'd like to present. I don't intend to prove either — although the proofs are interesting. The first concerns series of positive terms for which $a_{n+1} \leq a_n$. The above series converges if and only if

$$\sum_{n=1}^{\infty} 2^n a_{2^n}$$

converges. For those who've gone through a second semester of calculus, this replaces the Integral Test. In particular, if we look at the p-series — the series whose nth term is $1/n^p$, the above series becomes

$$\sum_{n=1}^{\infty} 2^n \frac{1}{2^{np}}$$

which is the geometric sum with ratio 2^{1-p}, which converges if $2^{1-p} < 1$, and diverges otherwise. So the p-series converges if $p > 1$ and diverges otherwise. I first saw this in Rudin's classic *Principles of Mathematical Analysis* [1], and he didn't name it (other than Theorem 3.27, or whatever), but to me it's the Thin Sub-Sequence Test because you're looking at a series whose terms are, well, a thin sub-sequence of the original sequence of terms.

One of the under-appreciated features of calculus is that while it is hard to evaluate sums of the form

$$\sum_{n=1}^{\infty} a_n$$

where the terms are positive and decreasing, it is often relatively easy to evaluate integrals — and some of those integrals can be used to obtain useful error estimates. If we have a decreasing function $f(x)$ such that $f(n) = a_n$, then the inequality

$$\int_{n-1}^{\infty} f(x)\,dx \geq \sum_{k=n}^{\infty} a_k$$

holds because the area under the curve $y = f(x)$ from $j - 1$ to j is greater than or equal to the area of the rectangle whose base is 1 and whose height is a_j. For instance, if we want to evaluate the p-series with $p = 3.5$ and have the error less than 0.0001, we are guaranteed of this as long as

$$\int_{n-1}^{\infty} x^{-3.5}\,dx = 0.4(n-1)^{-2.5} \leq 0.0001$$

This is easy to solve for n, we need $(n-1)^{2.5} \geq 4{,}000$, or $n-1 \geq 27.59$. So if we add up the first 29 terms of this series, it will be within 0.0001 of the sum. To five decimal places, the sum of the first 29 terms is 1.12665. I cranked up Excel and had it add the first 1,000 terms (I didn't want to overwork the poor thing), and got 1.126733855. By this point, it took 35 terms to change the 9th decimal place by 1, so the estimate looks pretty good to me.

Torricelli's Tower

Evangelista Torricelli was one of those 17th-century guys with multiple interests and contributions in multiple areas, but he was best known for his work in atmospheric science. He was the first to realize that winds resulted from a difference in temperature, and famously declared that "we live submerged in an ocean of air." He invented the barometer, and a unit of pressure, the torr, is named after him [2].

However, he also made notable contributions to mathematics. He was very interested in the development of infinite series, and a very famous solid, known as Torricelli's Trumpet, is named after him. The peculiar properties of this solid require integral calculus to appreciate, but the basic idea can be seen in the following construction, which deserves to be named Torricelli's Tower in his honor.

Torricelli's Tower consists of an infinite stack of cubes, one on top of another, and centered on a common axis, like cubes of sugar centered on a toothpick. You can rotate the cubes or line them all up so their faces are parallel according to your aesthetic sensibilities.

The bottom cube has side 1/1, the next cube has side $\frac{1}{\sqrt{2}}$, the third cube side $\frac{1}{\sqrt{3}}$, etc. Since the volume of each cube is the cube of the side, the sum of the volumes of all the cubes is

$$\sum_{n=1}^{\infty} \frac{1}{n^{3/2}}$$

As we know, this series converges. However, the nth cube has surface area $6/n$, and we know the series

$$\sum_{n=1}^{\infty} \frac{6}{n}$$

diverges. You may not be impressed — and I wasn't either, when I first was shown Torricelli's Trumpet — until my teacher pointed out that the fact that this object has finite volume but infinite surface area means you can fill it with paint but you can't paint it.

The next tool, which I'm tempted to prove but will resist temptation, is the Alternating Series Test. If we have a decreasing sequence of positive terms a_n, then the series

$$\sum_{n=1}^{\infty} (-1)^{n+1} a_n$$

converges (it's traditional to start an alternating series with a positive term so the sum is positive. What's seductive about the Alternating Series Test is that the error from adding up the first n terms is less than a_{n+1} (known as "the first neglected term," a description I've always enjoyed). For instance, for the series

$$\sum_{n=1}^{\infty} (-1)^{n+1} n^{-3.5}$$

the alternating p-series with $p = 3.5$, all we need to ensure an error less than 0.0001 is that $(n + 1)^{-3.5} < 0.0001$. So we need $(n + 1)^{3.5} >$ 10,000, and so $n + 1 > 13.9$. So 13 terms should be enough. Cranking up Excel once more, we see that the sum of the first 13 terms is 0.92761 to five decimal places, and the sum of the first 1,000 terms is 0.927553578. In this case, the ninth decimal place has remained the same for more than 550 terms!

But now it's time for what is undeniably one of the greatest theorems in mathematics.

Taylor's Theorem

There are other theorems that are more beautiful (I can think of several), and other theorems which are more profound (likewise). But if

I could have one theorem named after me — rather than the guy who it's named for — this is the one I'd choose.

Brook Taylor [3] was born to parents who belonged to one of the outer wings of British aristocracy, and was thus able to go to college and live a life of relative ease. Sadly, both of his wives died in childbirth, and only the daughter by his second wife survived. He did make contributions to several areas of calculus, including the discovery of integration by parts and the calculus of finite differences, but these are overshadowed by what is now called Taylor series, which first appeared in print in 1715.

And was basically ignored by the mathematical community until more than half a century later, when the brilliant French mathematician Joseph-Louis Lagrange provided an important contribution (which will be described shortly) and lauded it as "the main foundation of differential calculus." Praise from Lagrange is praise indeed.

So here's the idea. Assume that a function $f(x)$ can be written as a power series, which is just a polynomial with an infinite number of terms. Maybe a polynomial is restricted to having a finite number of terms (a quadratic has three, including the constant term, a cubic four, etc.), but let's suppose that

$$f(x) = a_0 + a_1 x + a_2 x^2 + a_3 x^3 + \cdots$$

Assuming that we can evaluate the function f at $x = 0$, by plugging in $x = 0$ on the left and right side, we quickly see that $a_0 = f(0)$. But how do we compute the values of the other coefficients?

This is easy — assuming you're familiar with first-semester calculus. The tool was the recently-developed differential calculus of functions of a single variable. If you differentiate both sides, assuming that series can be differentiated term by term as if they were polynomials (and why not?), we see that

$$f'(x) = a_1 + 2\, a_2 x + 3\, a_3 x^2 + \cdots$$

Plug in $x = 0$ to knock out all but the constant term on the right, and we see that $f'(0) = a_1$. Repeat the process to get $a_2 = f''(0)/2$. Keep doing it, and you'll see what Taylor saw, that $a_n = f^{(n)}(0)/n!$ Thus was born the famous Taylor series

$$f(x)=\sum_{n=0}^{\infty}\frac{f^{(n)}(0)}{n!}x^n$$

And if I could have one equation named after me, that would be the one. Props to Joseph-Louis Lagrange for recognizing how incredible this result was, and describing it as "the main foundation of differential calculus." And Lagrange's name is coupled with the series because he discovered a way to place an upper bound on the difference between the actual value of the function $f(x)$ and the sum of the first n terms of the series — it's known as the Lagrange form of the Remainder.

But it isn't necessary to know it in order to see how many terms of the series are needed to construct the four decimal place tables that occupied a prominent position in almost every math textbook printed in my high-school days. We only need to know the value of sin x for values of x between 0 and 90°.

Let's reprise the formula for sin x which blew me away when I first saw it.

$$\sin x = x - x^3/3! + x^5/5! - x^7/7! + \cdots$$

Since it's an alternating series, the Alternating Series Test will allow us to estimate the error (or remainder, if you like to look on the bright side) through the term $x^{(2n-1)}/(2n-1)!$ It's the absolute value of the first neglected term, which is $x^{(2+-1)}/(2n+1)!$ x has to be measured in radians in order that the derivative of $\sin x$ be $\cos x$, and not $\cos x$ with some ghastly coefficient like $\pi/180$ preceding it. Anyway, the largest x can be of the angles in a table in the back of my book is $90° = \pi/2$ radians, approximately 1.6. So all we need to do to have accurate four-place tables is for $1.6^{2n+1}/(2n+1)! < 0.0001/2$. This first happens when $2n + 1 = 11$, and so if we use the polynomial $x - x^3/3! + x^5/5! - x^7/7! + x^9/9!$ and use these values to construct the table of sines that appeared in my math books, you'll get the same numbers that they got.

This result cemented forever the value of infinite series — and, in fact, was the method of choice for calculating actual numerical values until about 150 years ago, when the mathematician Henri Padé noticed that in many cases, functions "blew up" near certain critical values — a behavior exhibited by rational functions (the name for quotients of polynomials). The best such functions, known as Padé approximants [4], can seriously improve upon the efficiency of computing accurate numerical values of functions — and that's something we definitely want to do. The practical definition of efficiency is the length of time a computer needs to calculate a numerical value of given accuracy — and even with the highest = speed computers, the less time you spend calculating meat-and-potatoes numbers, the more time you can spend calculating important stuff.

Playing with Infinite Series

One other important point about power series. We discussed earlier that the geometric series, which is a very specific power series, could be considered as a function, and if it were so considered, the domain of that function was an interval which we called the interval of convergence. Every power series has its own interval of convergence, and within that interval of convergence, we can differentiate it or integrate it as if it were a very long polynomial.

Here's an example with a fascinating history. Consider the series

$$1 - x^2 + x^4 - x^6 + x^8 - \cdots$$

Its sum is $1/(1 + x^2)$, and if we integrate this we obtain arc tan $x + C$ (the constant of integration). If we integrate the above series term by term, we get

$$x - x^3/3 + x^5/5 - x^7/7 + \cdots$$

Although the geometric series above does not converge for $x = 1$, the alternating series above does, so

$$\text{arc tan}\, x + C = x - x^3/3 + x^5/5 - x^7/7 + \cdots$$

When $x = 0$, both arc tan 0 and the right side are equal to 0, so $C = 0$ as well, and we have

$$\text{arc tan}\, x = x - x^3/3 + x^5/5 - x^7/7 + \cdots$$

In calculus we show that this result holds at $x = 1$, the edge of the interval of convergence of the original series. But arc tan $1 = \pi/4$, and thus we see that

$$\pi/4 = 1 - 1/3 + 1/5 - 1/7 + \cdots$$

I think this was the first formula that enabled π to be computed accurately to any number of decimal places — although the convergence is painfully slow. To compute π accurately to three decimal places, you need to be sure that the first neglected term is less than 5 ten thousandths, and that doesn't happen for a LONG long time. And remember, that first neglected term is actually the first neglected term in the series for $\pi/4$ (rather than π), so it takes even longer.

Offspring

Only one of Brook Taylor's children survived childbirth, and I don't know the Taylor family tree after that. But Taylor series spawned an enduring collection of progeny. Maybe 100 years after Taylor's inspiration, it occurred to Jean-Paul Fourier that instead of using polynomials as the building blocks for approximation, other families of functions might be more suitable. He chose initially to look at functions $f(x)$ which were periodic on the interval $[-\pi, \pi]$, and found the expansion

$$f(x) = \frac{a_0}{2} + \sum_{n=1}^{\infty} (a_n \cos nx + b_n \sin nx)$$

The expression on the right is known as a Fourier series, and if one truncates the series at some integer N, the expression is called a trigonometric polynomial.

Of course, the big question is — how do you get the coefficients? 100 years had passed since Taylor used differentiation to obtain the coefficients of his series, and it was time for integral calculus — the other branch of calculus — to make an appearance. The Fourier coefficients are obtained by integration according to the formulas

$$a_n = \int_{-\pi}^{\pi} f(x) \cos nx \, dx$$

$$b_n = \int_{-\pi}^{\pi} f(x) \sin nx \, dx$$

What joins Taylor series and Fourier series is the idea that functions can be approximated by sums of more recognizable functions — powers of x for Taylor series, sines and cosines for Fourier series. The coefficients for both series are found using the techniques of calculus.

Sines and cosines appear naturally in periodic phenomena — but also as solutions to the family of differential equations $y'' + n^2y = 0$. You can see this just by taking the second derivative of sin nx and plugging it into that equation, similarly with cos nx. Sturm-Liouville theory, developed in the second half of the 19th century, generalizes this. Families of differential equations give rise to solutions which serve as natural building blocks for certain functions, just as the sines and cosines are the natural building blocks for periodic functions.

Every so often, life synchs up with what I as a teacher am doing in the classroom, and it did so one day in the early 1980s when Los Angeles was shaken by an earthquake. The next day I was lecturing on a class of functions known as spherical harmonics — these functions emerge from a certain family of differential functions. Just as sines and cosines are used to model how strings vibrate, spherical harmonics are used to model how spheres vibrate — as the Earth does after an earthquake. As I recall, the students were VERY attentive the next day in class.

Riemann's Casino

We've discussed series which consisted of nothing but positive terms, and alternating series. If we take a look at the alternating series $1 - ½ + 1/3 - ¼ + \cdots$, we notice that the series itself converges, but if we replace all the minus signs by plus signs, we get the harmonic series (the p series with $p = 1$), which we know does not converge. Convergent series containing infinitely many positive and infinitely many negative terms, but which do not converge when all the minus signs are replaced by plus signs, are known as conditionally convergent series. And really weird things happen with conditionally convergent series.

The man who first noticed this was Bernhard Riemann, the son of a Lutheran pastor who showed up at the University of Gottingen with the intention of obtaining a degree in theology. But one of the things that could move Carl Friedrich Gauss ahead of Isaac Newton in the GOAT debate is that, unlike Newton, who basically did his own thing, Gauss helped foster the careers of other mathematicians. Theology's loss was mathematics' gain, as Gauss persuaded Riemann to pursue a career in mathematics. Despite dying shortly before his 40th birthday, Riemann made a number of extremely important contributions that we know of to several different branches of mathematics. He may also have made other contributions that we do not know of, as Riemann refused to publish incomplete works. Upon his death, his housekeeper discarded his unpublished works, which doubtless would have made the career of many lesser mathematicians [5].

One of Riemann's observations was on the weirdness of conditionally convergent series, which can be illustrated with an example I call Riemann's Casino. There are an infinite number of tables at Riemann's Casino, but the outcome of a single bet at each of the tables is predetermined. If you bet a dollar at table number $2n - 1$, where n is a positive integer, you lose a dollar. If you bet a dollar at table $2n$, where n is a positive integer, you win $1/n$ dollars.

Knowing that the p-series with $p = 1$ is divergent, you enter Riemann's Casino with the intention of betting a dollar at each even-numbered table — until a security guard calls your attention to a sign posted at the entrance of the Casino, which says, "Players must play once at each table."

This doesn't look good. If you go through the tables in numerical order, you lose a dollar at table 1, win a dollar at table 2, lose a dollar at table 3, win a half-dollar at table 4, lose a dollar at table 5, win a third of a dollar at table 6, etc. After every two tables, your bankroll takes a hit. But, as Riemann discovered, there is a way to beat the house.

After you have played at tables 1 and 2 (and broken even), you now skip to table 4 and play once at tables 4, 6, 8,..., 18, 20, 22. This gives you a total of 2.0198 dollars. You now go back to table 3 and fulfill your obligation to play there, losing a dollar. You now stick your 1.0198 dollars in your wallet and go to table 24, and play there, then at tables 26, 28,...., 168 and 170, where you win a total of 2.0058 dollars. You now go back to table 5 and fulfill your obligation to play there, losing a dollar. You now stick the 1.0058 dollars you have just won into your wallet, and observe that you have accumulated more than 2 dollars. So it's back to table 172 and the even-numbered tables, where you continue to play until you have won more than two dollars, and then a return to table 7 to fulfill your obligation to play at that table. By moving through the Casino in this fashion, you will eventually have won an infinite amount of money. You may think you have won all the money in the Universe, but not so — as we shall discuss in the chapter on infinity.

The result Riemann discovered is that the order in which you add up the terms in a conditionally convergent series matters so significantly that you can devise an order for adding up those terms so that the sum is any number you wish — including both plus and minus infinity! A truly fascinating — and totally unexpected — result.

John von Neumann and the Bumblebee

John von Neumann is described by Wikipedia as a Hungarian-American mathematician, computer scientist, physicist, economist and polymath. Well, duh, once we saw the first four nouns, we sort of guessed that he was a polymath. But von Neumann wasn't just content to sit at a desk and work on abstract problems, during World War II he played a key role in helping develop the atomic bomb as a member of the Manhattan Project. As a Hungarian émigré who had seen what happened when the forces of dictatorship run amok, he worried about Soviet superiority in the development of atomic weapons. Along with Edward Teller, he helped develop the H-bomb and the collateral theory of Mutual Assured Destruction, which has so far resulted in the detonation of only two atomic weapons during wartime — the ones used to end World War II [6].

I owe a personal debt to John von Neumann; the title of my doctoral dissertation was *Continuity of Homomorphisms of von Neumann Algebras* [7]. I'd be willing to bet that this is the first time in almost 50 years that this paper has been referenced, like most mathematics, its lifespan is a few years, after which mathematics has moved on. Of course, the great results of mathematics live on forever, but as I said, mathematics is an evolving edifice to which most mathematicians (including yours truly) make a few contributions which are then forgotten. The same can be said of most participants in any discipline, but the reason most people pursue something for a lifetime is because they get something out of it, not because they expect to be remembered. A lifetime spent in mathematics probably doesn't differ much from a lifetime spent in dance, except for the fact that it's probably easier to find employment in mathematics than dance.

Anyway, there is a very famous problem involving a bumblebee. Two trains start 200 miles apart, and each is moving towards the other on the same track at a constant speed of 50 miles per hour. A bumblebee starts at the front of one of the trains and flies at a constant speed of 20 miles per hour towards the other train. When it reaches the front of the other train, it turns around and flies at a constant speed of 20 miles per hour towards the first train. When it reaches the front of the first train, it turns around and flies at a constant speed of 20 miles per hour, and so on. How far does it fly before it is crushed to death between the two approaching trains?

This is a chapter on infinite series, and you can see how the flight of the bumblebee could be expressed as an infinite series of the sum of the distances it travels between turnarounds. But there is a really cute way to solve the problem without involving infinite series. The two trains are 200 miles apart and are approaching each other at a combined speed of 100 miles per hour, so they will collide in 2 hours. The bumblebee is flying at a constant speed of 20 miles per hour for those 2 hours, and so will travel 40 miles.

When this problem was given to John von Neumann, he took a minute to think about it and then gave the correct answer. His friend, who posed the problem, said, "That's correct, but I thought you'd sum the infinite series," — thinking that von Neumann had taken a while but had eventually seen the shortcut.

"You mean there's another way?" von Neumann responded.

It's a great story — but I'm not totally sold that von Neumann did set up and sum the infinite series — in his head — and did so under a minute. I've solved the problem in that fashion, just setting up the series (before summing) required pencil and paper.

But I do know this — REALLY brilliant people can think considerably deeper, and considerably more rapidly, than I can. So I'll give von Neumann the benefit of the doubt.

Bibliography

[1] See [2], Chapter 5.

[2] MacTutor. Online at https://mathshistory.st-andrews.ac.uk/Biographies/Torricelli/.

[3] MacTutor. Online at https://mathshistory.st-andrews.ac.uk/Biographies/Taylor/.

[4] Wikipedia. Online at https://en.wikipedia.org/wiki/Padé_approximant.

[5] MacTutor. Online at https://mathshistory.st-andrews.ac.uk/Biographies/Riemann/.

[6] MacTutor. Online at https://mathshistory.st-andrews.ac.uk/Biographies/Von_Neumann/.

[7] J. Stein, Continuity of Homomorphisms of Von Neumann Algebras, *American Journal of Mathematics*. Baltimore, MD, 1969.

Chapter 9

Seduced by Probability

If I had it to do all over again, when the time came for me to write a dissertation, I'd write it in the area of probability and statistics rather than the area in which I did write it, which is called functional analysis.

I didn't choose functional analysis, I chose Bill Bade, who was my professor in the analysis course I had to take as an entering graduate student, and who was the only teacher I ever encountered — in college or grad school — who gave lectures just the way I wanted them. So clearly were they presented that, even though I took copious notes, I could understand them in real time (a concept which did not exist while I was in grad school). Through what I still regard as a minor miracle, I passed my qualifying exams on the first try, I chanced to encounter him in the hall as I was leaving the exam room. I managed to persuade him to take me on as a doctoral student even though I hadn't the foggiest idea of what he did (but I hoped it wasn't algebraic topology, about which I understood *nada*). And I have no regrets about choosing Bill — and even though he isn't around to comment — I'm pretty sure he enjoyed having me as a doctoral student. After all, how many professors can tell their colleagues that one of their doctoral students has a lion in the basement? I'll save that story for a future book.

As you may have guessed, Bill's primary research area was functional analysis — and I did find it very interesting. But over the last few years, I have been completely obsessed with problems in probability with mind-blowing counterintuitive solutions — and fortunately, they can be understood with just a smidgen of high-school algebra.

Playing Games

Even though I have never taken a formal course in probability at any level, and even though I never even encountered the concept of probability in any of my high-school courses (back then, it just wasn't included in the curriculum as it is now), I was addicted to both bridge and backgammon, and learned as much elementary probability as I needed.

And that wasn't much. In bridge, all you need is a knowledge of the probability of how the missing cards in a suit figure to divide when split into two piles. For instance, if there are 5 missing spades, they will be divided 3–2 (3 in one pile, 2 in the other) 68% of the time, 4–1 28% of the time, and 5–0 4% of the time. In backgammon, you need to know some elementary facts about the probability of throwing a 1 with 2 dice (11 in 36), and either a 1 or a 2 with two dice (20 out of 36), or a 1 or a 2 or a 3 with two dice (27 out of 36), etc.

Finally, I spent 5 years trading stock options for a living, and trust me, when your money is on the line you learn a lot about expectation and reward/risk. So I did have some experience with probability — I'd guess I had the same level of familiarity with it as Girolamo Cardano [1], who invented the concept sometime in the 16th century. To be fair, I did teach some probability and statistics in a course entitled Finite Mathematics, which mostly existed to give college students a course that went beyond the high-school curriculum but didn't involve calculus.

I'd also been exposed to a few of the classic problems in probability — some of which you'll see in this chapter. But nothing about the subject caught fire with me until several years ago.

But first, let's look at a couple of deservedly well-known classic problems from probability.

The Birthday Problem

Because if there's any problem in mathematics that suffers from over-exposure, this is it, I'm going to modify the classic presentation of it.

You walk into a karaoke bar and are handed the list of songs, which contains one hundred well known favorites. Each participant gets to choose one and sing it. How many songs would you expect to hear before you hear the same song twice? What is being asked is how many songs must be chosen — on average — before you hear the same song twice?

The answer — 12 — is surprising. When told this problem, many people think that after half the songs — in this case, 50 — have been played, you'd expect to see the same one twice. But the correct way to calculate it is pretty easy. All you need to know is that the probability of two independent events (I'm assuming here that the choices of song are all made independently) occurring is the product of each occurring separately. As an example, the probability of a fair coin landing heads is ½. The probability of a random card drawn from a standard deck being a spade is ¼. So the probability of BOTH the coin landing heads and the card being a spade is $\frac{1}{2} \times \frac{1}{4} = 1/8$.

So the probability that the second song will be different from the first is 99 out of 100, 0.99. The probability that the third song differs from both the first and the second is 98 out of 100. So the probability of the first three songs all being different Is 0.99×0.98. The probability of the first 12 songs all being different is $0.99 \times 0.98 \times \cdots \times 0.89 =$ about 0.503, so that's where the dividing line is.

The classic problem has people with the same birthdays; the dividing line here turns out to be 23 people. If you have 23 randomly-chosen people, it's about 50–50 that two of them will have the same birthday.

The Monty Hall Problem

Monty Hall was the host of *Let's Make a Deal* [2], a TV show that first appeared in 1960s got canceled, and then made a reappearance in the 21st century. I don't know who originally described the problem, but

Marilyn vos Savant wrote about it in her column *Ask Marilyn* [3] a number of years ago, and said she got twice as many letters on this column as on any other. You can guess how long ago this was from the word "letters" in the previous sentence.

A contestant on *Let's Make a Deal* has earned the chance to win a prize of $50,000 — which lies behind one of three numbered doors on the stage. Behind the other two doors is a bar of the sponsor's soap. Monty Hall asks the contestant to select a door, and she will receive what's behind the door — either cash or soap.

Whenever I describe this problem to someone, I ask them to choose a door. For some absolutely unknown reason, almost all of them choose Door #2. So let's assume you do as well.

Monty now walks over to the three doors, and opens Door #1. Behind it — not surprisingly — is a bar of soap. This isn't surprising for two reasons. First, if Monty opened the door behind which the cash was located, all the suspense would be gone — and Monty would soon have to update his resumé, as he would shortly be the former host of *Let's Make a Deal*. Second, the sponsor clearly welcomes more air time for his product.

Monty now turns to the contestant and asks her, "Would you like to stay with your choice, Door #2, or switch to Door #3?" Well, what would you do?

There are several variations of this problem, in which Monty incentivizes either staying or switching, but let's suppose we go with the vanilla version described above. Do you stay or switch — and why?

At this point, many people take one of two positions. The first group of people believe that it's fifty-fifty — the prize is obviously behind Door #1 or Door #3, and one door is the same as any other door. The second believe that there's something sinister afoot — they guessed correctly, the sponsor doesn't want to lose $50,000, and Monty is trying to talk them out of the prize. But the answer, though initially counterintuitive, has a very straightforward explanation.

You should switch your choice from Door #2 to Door #3, because the cash is twice as likely to be behind Door #3! This is so counterintuitive that it explains why Marilyn vos Savant got a huge number of letters about it. But the idea is quite simple — when you chose Door #2, you divided the doors into two groups — your door and the remaining doors. There are two remaining doors (Doors #1 and #3), so the prize is twice as likely to be behind one of them as behind Door #2.

Some people are a little dubious about this argument. Remember in Chapter 1, when we looked at an extreme situation in the case of the bullet holes in the airplane to make the right answer clear? We'll reprise that idea here. Let's assume there aren't three doors, but a thousand. Monty now offers you the choice of sticking with Door #2 (again you chose #2), or taking the cash if it's behind any one of the other 999 doors. Are you really going to stick with Door #2, or switch in the hope that the cash is behind one of those other 999 doors? And when you decide to switch, if he opens all the doors except Door #3 and offers you the chance to switch back to Door #2, it's almost certain that you wouldn't do that. Did you really think you were so lucky that, out of one thousand doors, you picked the door with the cash?

Bayes' Theorem

Bayes' Theorem is of tremendous importance, both in the natural sciences, the social sciences, and everyday life. Surprisingly, it is so easy that even an elementary school student can do problems involving Bayes' Theorem — as you'll soon see.

Let's suppose that there's a town in which 20% of the people are left-handed and the other 80% are right-handed. 70% of the lefties like asparagus, as do 60% of the righties.

Question 1 — What percentage of the town likes asparagus?

Question 2 — What percentage of people who like asparagus are left-handed?

Question 3 — What percentage of people who don't like asparagus are right-handed?

Every single one of these questions can be answered via simple arithmetic — REALLY simple arithmetic. Let's imagine the town has 100 inhabitants. Of those, 20 are left-handed and 80 are right-handed. Of the 20 lefties, 70% like asparagus, that's 14 people, and the other 6 don't. Of the 80 righties, 60% like asparagus, that's 48 people, and the other 32 don't. So the town breaks down into four groups as follows.

Handedness	Feelings about Asparagus	Number
Left	Like	14
Left	Don't like	6
Right	Like	48
Right	Don't Like	32

We can use these to answer all three questions very quickly.

Q1: $14 + 48 = 62$ of 100 people like asparagus, or 62%

Q2: There are 62 people who like asparagus. 14 of them are left-handed. So the percentage of people who like asparagus that are left-handed is $100 \times 14/62$, or 22.58%.

Q3: There are 38 people who don't like asparagus. 32 of them are right-handed. So the percentage of people who don't like asparagus that are right-handed is $100 \times 32/38$, or 84.21%.

Notice that the information is presented originally in the following order — handedness percentage first, then feelings about asparagus once handedness has been determined. The percentage of lefties who like asparagus is known as a conditional probability — we're asking whether someone who satisfies the condition that they are left-handed likes asparagus.

Questions 2 and 3 reverse the order — we're asking about handedness percentage subject to the condition of feelings about asparagus. It is this "reverse order" that characterizes Bayes' Theorem.

I used an example involving handedness and feelings about asparagus because neither is a hot-button issue. Nobody likes or dislikes anyone because they are left-handed or because they don't like asparagus. But change the environment to political views on hot-button issues — or on how to market your product — and Bayes' Theorem becomes extremely important.

In particular, Bayes' Theorem is important to paternity testing via DNA analysis. This subject is not easily understood by many jurors, and expert witnesses are frequently needed, not only to "do the math" but to explain how the math affects a particular case.

Tennis, Anyone?

I think this question was first posed in this format in some drawing-room comedy in the 1920s or 1930s, when some appropriately-garbed character wanders into the aforementioned drawing room clutching a tennis racket and uttered that phrase. But, had I been in that room, I would have answered affirmatively, because I've been a tennis devotee for more than half a century.

Tennis scoring is similar to scoring in table tennis (not surprisingly) and volleyball — if the score is tied after a certain number of points, in order to win the game a player must win by a margin of two points. Tennis, however, differs from volleyball and table tennis in that the same player retains the serve (which starts the point) until the completion of the game. In tennis, if each player has won three points, the score is known as "deuce" (no idea why). So an obvious question (for the mathematically-inclined, such as me) is — if the player who serves has a probability p of winning each point, what is the probability that he (or she) will win from deuce?

It's not hard to see that this problem can be solved via summing an infinite series. Assuming that the letter W stands for "server wins the next two points" and D stands for "server and returner each win one of the next two points," the different ways that server can win are W, DW (one deuce, then server wins), DDW (two deuces, then server wins), $DDDW$,... and so on.

The probability of the server winning the next two points is p^2. The probability that the server and receiver alternate winning the next two points is $p(1 - p) + (1 - p)p = 2p(1 - p)$. The probability that the game will return to deuce k times and then the server will win the next two points is therefore seen to be $(2p(1 - p))^k p^2$, and when we sum this infinite series we see that the probability of the server winning from deuce is $p^2/(1 - 2p(1 - p))$.

This solution is fairly straightforward, but there's an alternative way to do this which is considerably more attractive. Let P denote the probability that the server wins from deuce. If two points are played, there are three possible outcomes. The first (sadly, from the server's POV) is that receiver wins the next two points and ends the game. The second (happily, from the server's POV) is that server wins the next two points and emerges victorious, this happens with probability p^2. The third is that the outcome is still undecided — server and receiver alternate winning the next two points, and we're back at deuce. This third possibility happens with probability $2p(1 - p)$, as we saw in the previous paragraph. Therefore, P is the sum of the probabilities that the server wins the next two points (this is p^2) or the game returns to deuce (which happens with probability $2p(1 - p)$) and the server THEN wins from deuce. This results in the equation $P = p^2 + 2p(1 - p)P$, the solution to which is easily seen to be $p^2/(1 - 2p(1 - p))$.

Tennish, Anyone?

I first approached the problem of winning from deuce when the world and I were young, sometime back in the 20th century. But recently Len Wapner, a friend and fellow author, communicated the following problem to me.

Imagine you and I play a single, tennis-like game. (The game only loosely resembles a tiebreaker and is not intended to model one.) Assume we are of equal ability and the serve will change to the other player after each point is played. Let's assume the server (be it you or I) for any single point has a probability p of winning the point. The winner of the game is defined as the first to win two consecutive points. Here are two such games.

Game 1
Len wins the point.
Len wins the point.
Game over. Len wins the game.

Game 2
Jim wins the point.
Len wins the point.
Jim wins the point.
Len wins the point.
Jim wins the point.
Jim wins the point.
Game over. Jim wins the game.

What is the probability that the first person to serve will win the game?

This was clearly a more complicated problem than winning from deuce, and I really didn't want to get bogged down in an analysis which required me to categorize and sum a number of infinite series. So I wondered if the probabilistic method used in winning from deuce might apply. A few hours — and a number of false starts — later, I came up with the following solution.

Let A serve first, P the probability of A winning, p the probability of the server winning the point.

The game has four "live" states where the game has not been decided.

State 00 — A serves, A is up one point

State 01 — A serves, B is up one point

State 10 — B serves, A is up one point

State 11 — B serves, B is up one point

Let Pij denote the probability that A wins from state ij. We have the following five equations.

(1) $P = pP10 + (1 - p)P11$

If A wins the point, we're in state 10. If B wins the point, we're in state 11.

(2) $P00 = p + (1 - p)P11$

If A wins the point, A wins. If B wins the point, we're in state 11.

(3) $P01 = pP10$

If A wins the point, we're in state 10. If B wins the point, A loses.

(4) $P10 = pP01 + (1 - p)$

If A wins the point, he wins. If B wins the point, we're in state 01.

(5) $P11 = (1 - p)P00$

If A wins the point, we're in state 00. If B wins the point, A loses.

The solution to (3) and (4) is $P10 = 1/(1 + p)$. The solution to (2) and (5) is $P11 = (1 - p)/(2 - p)$. Plugging them into (1) gives $P = p/(1 + p) + (1 - p)^2/(2 - p)$.

If you either graph this or do some calculus, the result is counter-intuitive. If the probability of the server winning is less than 0.5, the server is the favorite, but if the probability of the server winning is greater than 0.5, the receiver is the favorite. This is rather surprising — if serving is advantageous on a point, it is disadvantageous in a game — and vice versa!

Math never ceases to surprise and delight me.

Entanglement in Quantum Mechanics

Entanglement generally means something like "mixed together and difficult to separate," but the everyday definition was taken to a totally different level by quantum mechanics. So let's take a look at what entanglement means in what is possibly the most subtle and perplexing example of behavior in the subatomic world.

Subatomic particles have a property called "spin" — physicists tell us that it's not really spin as we envision it (neither is spin in the world of politics, for that matter), but that's the word they've used to describe it. An individual particle can spin to the left or to the right, but we'll imagine a system in which two entangled particles have a net spin of zero — when spin is measured, one will spin to the left, the other to the right.

That's not all that hard to accept — but it gets weirder. MUCH weirder.

The spin of each of our entangled particles is a probability distribution — half the time it spins to the left, half the time to the right. This isn't that hard to accept, either — if we flip a fair coin in the air, while it is still spinning (in the conventional sense of the word), the eventual landing state of the coin is also a probability distribution — half the time it will land heads, half the time it will land tails. We don't know which way each of our entangled particles spins until we measure it. And that isn't that hard to accept, either — we don't know whether the eventual landing state of the coin is heads or tails until — duh — it lands.

But here's the example that had Einstein rolling his eyes in disbelief. Suppose that we separate the two entangled particles by moving them billions of light years apart. We then measure the spin of one of them, and discover that it spins to the left. Instantly, the other entangled particle billions of light years away "gets the memo" and knows that it is to spin to the right. And it does — every time. After more than a century, physicists know **that** this is done — but they have no idea **how** it is done. Einstein described this as "spooky action at a distance." It puzzled him, and a century later, it remains a puzzle.

You might decide this is such a cool phenomenon that you want to investigate it. Fortunately, you can do so in your own home by purchasing a photon entangler. You won't find it by that name, but if you look under "spontaneous parametric down conversion" — the physicists' name for photon entangler — you can find tabletop versions. They're made of lasers and computers — and will probably put a dent in your bank account. However, they can do some really neat things — and may become a staple device for quantum encryption, which may soon prove necessary.

Probabilistic Entanglement

However, spontaneous parametric down conversion is not so useful for everyday practical problems. Let's say you're running a company which is manufacturing a device, and you have a choice of two equally-priced components of unknown reliability. You only have time enough — or budget enough — to make one test, and then you've got to choose. Although it may seem like a 50–50 proposition, there is a way to improve those odds — by use of a forensically advantageous interference resistant chaotically oriented inference normalizer — best known by its acronym, FAIRCOIN. You've probably got one lying around the house, and they cost next to nothing.

So here's how you use FAIRCOIN to improve your odds of making the correct decision between Component A and Component B. Flip FAIRCOIN — if it lands heads, test Component A and if it lands tails, test Component B. If the test results in a success, select that component for use, and if it results in failure, select the other component.

Here's the math. Let's suppose that Component A has success probability p, Component B has success probability q, and we'll assume that $p > q$ (although this is not known to you when you make the test). Through the use of FAIRCOIN, the probability of testing each component is ½. If you test A and it is a success, you'll make the correct decision — the combined probability of testing A and having the test be successful is $\frac{1}{2} p$. If you test B and it is a failure, you'll make the correct decision — the combined probability of testing B and having the test result in failure is $\frac{1}{2}(1 - q)$. So the probability that you'll make the correct decision is $\frac{1}{2} p + \frac{1}{2}(1 - q) = \frac{1}{2} + \frac{1}{2}(p - q)$, which is greater than ½ and could be considerably greater if $p - q$ is sizeable.

I've coined (play on words intentional) the phrase "probabilistic entanglement" to describe the above process because it is much more far-reaching than the example above might suggest. FAIRCOIN can enable you to guess correctly more than 50% of the time which of two phenomena is more likely to occur, no matter how unrelated the phenomena may be.

Here's an example. I live in Redondo Beach, California — this book is being published in Singapore. Suppose we want to know which percentage is larger, the percentage of people in Redondo Beach with last names of five letters or less, or the percentage of people in Singapore who own a pet. Get out your FAIRCOIN and flip it. If it lands heads, pick a random person in Redondo Beach and count the letters in their last name — it's a success if it's five or fewer letters, a failure otherwise. If FAIRCOIN lands tails, select a random person in Singapore and find out if they own a pet — it's a success if they do, a failure if they don't. If the test resulted in a success, guess that the population from which the test was made exhibits the larger percentage; and if the test resulted in a failure, guess that the other population has the larger percentage. The math is exactly the same as with the components.

Since I live in Redondo Beach, it would obviously be simpler for me to decide that I don't want to go through the effort of locating a random person in Singapore and asking whether they have a pet. It's easier for me just to find a random person in Redondo Beach and count the number of letters in their last name. If there are five or fewer letters, I guess that the percentage of people in Redondo Beach with short last names is greater than the percentage of people in Singapore who own pets; if there are six or more letters, I guess that the pet-owning percentage in Singapore is greater.

Let's suppose that the probability of a random person in Redondo Beach having a last name of five or fewer letters is p, and the probability of a random person in Singapore owning a pet is q. If $p > q$, I'll make the correct decision with probability p — but if $p < q$, I'll make the correct decision with probability $1 - p$. In the absence of any information, each appears equally likely, so I'll make the correct decision with probability $\frac{1}{2}\, p + \frac{1}{2}\,(1 - p) = \frac{1}{2}$. Even if we don't make the "each is equally likely" assumption — and make no assumption whatsoever — we'll be right with probability p and wrong with probability $1 - p$ — and if $p < \frac{1}{2}$, we'll be wrong more often than not.

What FAIRCOIN does is "probabilistically entangle" the two disparate data sets. Like the two photons that are billions of light years apart but are still entangled, FAIRCOIN entangles ANY two data sets, no matter how many billions of light years of distance and eons of time separate them. Even if we end up testing someone in Redondo Beach, there has to be the chance that we could have tested someone in Singapore.

Blackwell's Bet

I'm concluding this chapter with a truly remarkable discovery made by David Blackwell [4] — a truly remarkable individual.

Thanks to the book and movie *Hidden Figures* [5], many people know the story of Katherine Johnson, an amazing woman who worked as a computer for NASA, starting in the era when "computer" was a job description rather than a piece of household equipment. Katherine Johnson was a black woman born in 1918 who managed to overcome both of these obstacles to play an important role in the American space program of the 1960s. If you haven't either read the book or seen the movie, you won't be disappointed by taking the time to do either — or both.

David Blackwell was born in 1919, and faced only one obstacle — being black. Nonetheless, his mathematical brilliance surfaced early, and he earned a fellowship to Princeton, where he was denied the right to attend lectures because he was black. He was interested in statistics, and applied to the University of California at Berkeley in the 1940s for a teaching position. His appeal was supported by Jerzy Neyman, one of the pre-eminent statisticians of the era, but he was initially rejected because it was the tradition of the department chair to invite new hires to his home for dinner, and the department chair felt (or knew) that his wife would be uncomfortable inviting a black man to dinner.

A decade later, attitudes changed — and Blackwell was then hired at Berkeley. This proved to be a great move, as Blackwell eventually became the first black to be admitted to the National Academy of Sciences. He was universally beloved, both by students and his colleagues.

I was first acquainted with the scenario I'm about to describe by Len Wapner, who named it Blackwell's Bet. It is one of the most counterintuitive results in mathematics that I've ever encountered — and the math underlying it is really simple.

Let's suppose there are a white envelope and a brown envelope, each containing a different amount of money. You choose one of these two envelopes at random, open it, and count the amount of money in it. Like the Monty Hall problem, you are now offered the opportunity to either keep the amount of money in the envelope you just opened, or switch and take the amount of money in the unopened envelope. Because the value of the money to you (what economists call *utility*) might play a role — for instance, you'd almost certainly keep the amount of money if it were a million dollars and you'd almost certainly switch if the amount of money were a trivial amount such as one cent), let's just assume that your goal is simply to choose the correct envelope. Is there anything you can do to improve what looks like a 50–50 proposition?

Surprisingly enough, there is — and even though I have more than half a century of mathematical experience, when I was presented with this problem the solution never occurred to me. So I'll give you until the next paragraph to think about it.

Here's how you improve your chances. Pick a random number from anywhere — the current balance in your checkbook will serve as a good example. Compare that random number with the amount of money in the envelope you just opened. If the random number is less than the amount of money in the opened envelope, keep the money. If the random number is greater than the amount of money in the opened envelope, switch to the other envelope.

Startling as this may be — and it blew me away when I was told the solution — the math is simple and straightforward. Suppose the smaller amount of money is S, and the larger amount of money is L. When you initially choose an envelope — before you opened it — you were equally likely to have chosen the envelope containing S as the envelope containing L.

Now let p be the probability that the random number you chose is less than S, and q the probability that the random number you chose is greater than L. If you opened the envelope containing S — which you did with probability ½, the probability that you will make the correct decision is $1 - p$. The combined probability of choosing the envelope containing S AND making the correct decision is therefore ½ $(1 - p)$.

Similarly, if you opened the envelope containing L — which you did with probability ½, the probability that you will make the correct decision is $1 - q$. The combined probability of choosing the envelope containing L AND making the correct decision is therefore ½ $(1 - q)$.

Therefore, the probability that you will make the correct decision is

$$\tfrac{1}{2}(1 - p) + \tfrac{1}{2}(1 - q) = \tfrac{1}{2} + \tfrac{1}{2}(1 - (p + q))$$

The last term on the right is half the probability that the random number you chose falls in the gap between S and L — and this will happen some of the time. Therefore, this method will give you better than an even chance of making the correct decision.

I found this result so seductive that I not only spent several years learning about it and thinking about it, I also (shameless plug) wrote a book about it — *The Fate of Schrodinger's Cat* [6]. Since that book is published by the same firm that is publishing this book, they probably won't object to me mentioning it.

I received my doctorate from the University of California at Berkeley under the guidance of Prof. William Bade, a brilliant, kindly, and understanding mentor. I wouldn't change anything about that experience. But I'm a believer in the many-worlds theory of cosmology, and I think that there are universes in which somehow, while I was at Berkeley, I took a course in either probability or statistics from David Blackwell, who was there at the same time I was. And I hope he enjoyed having me as a student as much as I enjoyed having him as an instructor.

Bibliography

[1] MacTutor. Online at https://mathshistory.st-andrews.ac.uk/Biographies/Cardan/.
[2] Wikipedia. Online at https://en.wikipedia.org/wiki/Let%27s_Make_a_Deal.
[3] Jstor. Online at https://www.jstor.org/stable/3619484.
[4] MacTutor. Online at https://mathshistory.st-andrews.ac.uk/Biographies/Blackwell/.
[5] Wikipedia. Online at https://en.wikipedia.org/wiki/Hidden_Figures.
[6] J. Stein, *The Fate of Schrodinger's Cat*. Singapore: World Scientific Publishing, 2020.

Chapter 10

Seduced by Infinity

One of my favorite movies is *2001: A Space Odyssey*. Using Arthur C. Clarke's classic short story *The Sentinel* [1] as its jumping-off point, the movie is a technological and visual *tour de force*. The climactic final scene, which has been analyzed by film critics, psychologists and pop sociologists, begins with the line, "Jupiter — and beyond, the infinite."

Well, the infinite is certainly beyond Jupiter — way beyond. For millennia philosophers had tried to come to grips with the infinite — mostly unsuccessfully, IMHO. Mathematicians approached the subject of infinity gingerly, as something that could be approached, but not attained. Until Georg Cantor [2], that is.

History — even mathematical history — goes through revisionist phases. Had I written this section twenty years ago, I would have made a parallel between Georg Cantor's life and the life of Vincent van Gogh. Twenty years ago, there was a tendency to view Cantor as a brilliant outcast of the mathematical community, driven to depression and religious mania by the rejection of his work. In fact, for the most part, Cantor was held in extraordinarily high esteem by the mathematical community. True, some of his work was not immediately accepted — in 1884, one eminent journal said that publication of one of his papers would be 100 years too early. But his work in mainstream mathematics was not only exceptional but was accepted as such by the mathematical community. David Hilbert, one of the towering figures of 20th century mathematics, described Cantor's work as "the finest product of mathematical genius and one of the supreme achievements of purely intellectual human activity."

Among Cantor's achievements were the creation of set theory and the theory of transfinite numbers, some of which we'll touch on here. But I look at Cantor as one of the few mathematicians who has theorems named after him (Cantor's Theorem in set theory), objects named after him (the Cantor set in the real line), and a method of proof named after him (Cantor's diagonal process). Few mathematicians have hit for this particular cycle.

So let's take our first look beyond Jupiter, at the infinite.

Sleight of Hand

Somehow, you find yourself in possession of a collection of numbered balls. It's a big collection, there's one ball in the collection for each positive integer n. Your mission, should you decide to accept it, is to transfer balls into a jar.

At 11:00 AM, you dump balls 1 through 10 into the jar, and remove ball 1.

At 11:30 AM, you dump balls 11 through 20 into the jar, and remove ball 2.

At 11:45 AM, you dump balls 21 through 30 into the jar, and remove ball 3.

In general, at $1/2^k$ hours before noon, you dump balls $10k + 1$ through $10(k + 1)$ into the jar, and remove ball $k + 1$.

OK, you're going to be REALLY busy during the last second before noon, but you are up to the task. The clock strikes noon. How many balls are in the jar?

Your first thought — and I wouldn't blame you at all for thinking this — is that there are an infinite number of balls in the jar. After all, every combination of "dump and remove" adds nine balls to the jar. You're doing this an infinite number of times — so there should be an infinite number of balls in the jar. Not an unreasonable thought.

But here's the problem — can you name a single ball in the jar? Ball 1 isn't in the jar, you removed it at 11:00 AM. Ball 2 isn't in the jar, you removed it at 11:30 AM. Ball 3 isn't in the jar, you removed it at 11:45 AM. In fact, ball number k was removed $1/2^{k-1}$ hours before noon, so the correct answer is that no balls are in the jar.

But how can that be? Every single "dump and remove" added nine balls to the jar, and there are no balls left? Are you kidding?

And that's why I titled this section **Sleight of Hand**. Every time you went into action, you added nine balls to the jar — but when noon came and you looked at the jar, all the balls were gone. If that isn't major-league sleight of hand, I don't know what is.

You're just starting to come to grips with some of the apparent paradoxes that made mathematicians nervous about messing with infinity.

Let's change the "dump and remove" policy a little. Let's rewrite the general rule slightly, and say that at $1/2^k$ hours before noon, you dump balls $10k + 1$ through $10(k + 1)$ into the jar, and remove ball $2k + 1$. With this rule, you are removing all the odd-numbered balls, and when you examine the jar at noon, you'll find that it does indeed contain an infinite number of balls, as all the even-numbered balls are still in the jar.

In fact, by rewriting the general rule appropriately, we can end up with ANY number of balls in the jar at noon. Are we really to conclude that $9 + 9 + 9 + \cdots =$ whatever?

We actually touched on a couple of problems involved with adding up an infinite collection in the chapter on infinite series. The rules of ordinary arithmetic do apply to convergent infinite series, but when you leave the realm of convergent infinite series, other rules are needed. It wasn't until Georg Cantor came along that mathematicians actually got it right.

One-to-One Correspondences

One of the courses I enjoyed teaching was Math for Elementary School Teachers, because I got to say such things as, "I've been teaching math for thirty years, and I really can't tell you what 'three' is." When we're in elementary school, we are presented with a picture of three cookies and a picture of three oranges, and asked to connect each cookie with an orange, and each orange with a cookie, by drawing connecting lines. Then we do it with three oranges and three stars, or three houses. In doing so we learn the essence of "three-ness" — the term I use for the prospective elementary school teachers. We may not know exactly what three is — other than to say it's that ineffable something that collection of three cookies, three oranges, and three stars have in common.

But what's really critical here, and what was Cantor's first insight, was that the key idea was the one-to-one correspondence. The connecting lines established a one-to-one correspondence between cookies and oranges. And it took Cantor to realize that this was the crucial element in coming to grips with infinity.

Finite and Countable Sets: Hilbert's Hotel

It probably won't surprise you to know that a finite set is a set that, for some positive integer N, can be put in one-to-one correspondence with the integers 1, 2,..., N. Instead of putting three cookies in one-to-one correspondence with a set consisting of three oranges, we put it in one-to-one correspondence with the set (1,2,3). And that's how we learn to count.

The first move in dealing with infinity is to realize that there is a convenient infinite set, the set of ALL positive integers. We say that a set is countable if we can put it in one-to-one correspondence with the set of all positive integers.

David Hilbert came up with a really good way to illustrate some of the peculiar properties of countable sets. He envisioned a hotel with an infinite number of rooms, one for each positive integer. I don't think he called it Hilbert's Hotel, but everyone else did.

One night Hilbert's Hotel was doing especially good business, and every room was full. Then another guest showed up and asked if there was a room available. Despite the fact that every room was full, there's always room for another guest at Hilbert's Hotel. Hilbert moved the guest currently occupying Room 1 to Room 2, the guest currently occupying Room 2 to Room 3, the guest currently occupying Room 3 to Room 4, and so on. After he finished doing so, Room 1 was now vacant — and that's where the new guest went.

The next night the hotel was still full, and countably many new guests showed up. This required a little more ingenuity — but Hilbert's Hotel can accommodate countably many new guests, even though it is full! Hilbert moved the guest currently occupying Room 1 to Room 2, the guest currently occupying Room 2 to Room 4, the guest currently occupying Room 3 to Room 6, and so on. After he finished doing this, all the odd-numbered rooms were vacant, so Hilbert gave Room 1 to the first new guest, Room 3 to the second new guest, Room 5 to the third new guest, and so on.

You're probably asking — as were the mathematicians of the day — could Hilbert's Hotel accommodate ANY collection of new guests? And it took the brilliance of Georg Cantor to answer it.

An Influx of Bynars

David Hilbert died in 1943, but thanks to one of those unusual temporal anomalies that show up on *Star Trek* and its myriad spinoffs, one day an influx of Bynars descended upon Hilbert's Hotel, which had been completely emptied for spring cleaning. The Bynars do not have names but numbers that are constructed from the digits 0 and 1. In the episode featuring the Bynars [3], the numbers are finite in length, such as 10110101, but the Bynars reproduce so rapidly that each Bynar seeking accommodations in Hilbert's Hotel has a countable sequence of 0s and 1s as its designation. Bynar 101010...., for example, has 1 as the digit in the first, third, fifth, and every odd position, and 0 in the second, fourth, sixth, and every even position. The Bynars that sought rooms in Hilbert's Hotel had as designations every possible such sequence of 0s and 1s.

Try as we might, we will be unable to devise a scheme for assigning rooms in Hilbert's Hotel to all the Bynars clamoring for accommodations. And, in a brilliant proof, Georg Cantor showed why this was not possible.

Suppose, said Cantor, we had a scheme which assigned every Bynar to a room. There was a Bynar in Room 1, one in Room 2, one in Room 3, etc. Cantor was able to produce a Bynar that had no room — and he did it through a brilliant process now known as a Cantor diagonal proof.

Let $D(k)$ denote the designation (the sequence of 0s and 1s) of the Bynar in Room k, and let $D_n(k)$ denote the nth digit of $D(k)$ — recall that each $D(k)$ has a first digit, a second digit, a third digit, and so on forever. Now construct the designation X whose nth digit is $1 - D_n(n)$. In case this isn't immediately clear, what we've done is looked at the nth digit of the Bynar in Room n, and made sure that the nth digit of X is 1 if the nth digit of the Bynar in Room n is 0, and the nth digit of X is 0 if the nth digit of the Bynar in Room n is 1.

The diagram below shows how the designation for X is constructed.

Room 1	①	0	1	1	...
Room 2	0	⓪	1	0	...
Room 3	1	1	⓪	0	...
Room 4	0	1	1	⓪	...

The Bynar in Room 1 has designation 1011... . The first digit of this designation is 1, so the first digit of X is 0.

The Bynar in Room 2 has designation 0010... . The second digit of this designation is 0, so the second digit of X is 1.

The Bynar in Room 3 has designation 1100... . The third digit of this designation is 0, so the third digit of X is 1.

And so on. I've circled the nth digit of the nth designation to make them stand out, and so you can see why this is known as a Cantor diagonal proof. By construction, the nth digit of X is different from the nth digit of the designation of the inhabitant of Room n, and so X has not been assigned to any room. This contradicts the assumption that there is such a scheme of assigning rooms to Bynars so that each Bynar is the sole occupant of a room.

I don't think this was Cantor's original proof (he certainly didn't use Hilbert's Hotel or Bynars), but the idea behind the proof is his. And there's an immediate important consequence. If you put a decimal point in front of each Bynar designation, you have the binary representation of every real number x between 0 and 1. For example, the Bynar in Room 1 would be the real number that starts ½ + 1/8 + 1/16 + ⋯. Therefore the set of all real numbers between 0 and 1 (this set is known by the awe-inspiring name *the continuum*) could not be put in one-to-one correspondence with the integers. The real numbers were a larger set than the integers in a fashion that could be quantified, via cardinal numbers.

Cantor's Theorem

Cantor took this one step further. Using the diagonal process he had developed, he showed that there was literally no end to the cardinal numbers.

There are several different ways of stating Cantor's Theorem. The usual one is that the set of all subsets of a set X has a larger cardinality than the cardinality of the set S. In other words, the set of all subsets of X cannot be put in one-to-one correspondence with X.

Cantor's proof of this theorem is the essence of seductive simplicity. Let φ denote the set of all subsets of X, and suppose there is a one-to-one map $f: X \rightarrow \varphi$. In other words, for each x in X, $f(x)$ is a subset of X. Playing the role of the roomless Bynar here will be a set that we will construct that is not an $f(x)$ for any x in X. Let S be the set of all x in X such that x does not belong to $f(x)$.

Suppose that there was an element u in X such that $S = f(u)$. If u belongs to S, then since $S = f(u)$, u belongs to $f(u)$. But by definition, S is the set of all x in X such that x does not belong to $f(x)$, and so if u belongs to S, u does not belong to $f(u)$. We can't have u both belonging to $f(u)$ and not belonging to $f(u)$, so u cannot belong to S.

But if u does not belong to S, since $S = f(u)$, u does not belong to $f(u)$. But since u does not belong to S, it is false that u does not belong to $f(u)$, therefore u belongs to $f(u)$, and we end up with the same contradiction as in the last paragraph. Since both assumptions — u belongs to S and u does not belong to S — lead to contradictions, there cannot be a u in X such that $f(u)$ = S. Just like with the Bynar who had no room in Hilbert's Hotel, we have found that it is impossible to have a one-to-one correspondence between X and the set of all subsets of X.

The Continuum Hypothesis

Cantor established that the continuum — the set of all real numbers in the closed interval [0, 1] — could not be put in one-to-one correspondence with the integers. Cantor denoted the cardinality of the integers by aleph-0, aleph being the first letter of the Hebrew alphabet, and the cardinality of the continuum by aleph-1. Cantor was also able to show that the cardinality of the rationals was aleph-0, as was the cardinality of the algebraic numbers (the set of all real solutions to polynomials with integer coefficients). This was done using another diagonal argument. I'm not sure how Cantor did it, but proving that a countable union of countable sets is itself countable will take care of it; since part of the purpose of this book is to give proofs that I (and other mathematicians) find seductive, I'll do it for this one.

Suppose that the set $A_n = \{a_{nk} | k = 1, 2, \ldots\}$. Arrange the union of the A_n in a matrix as follows.

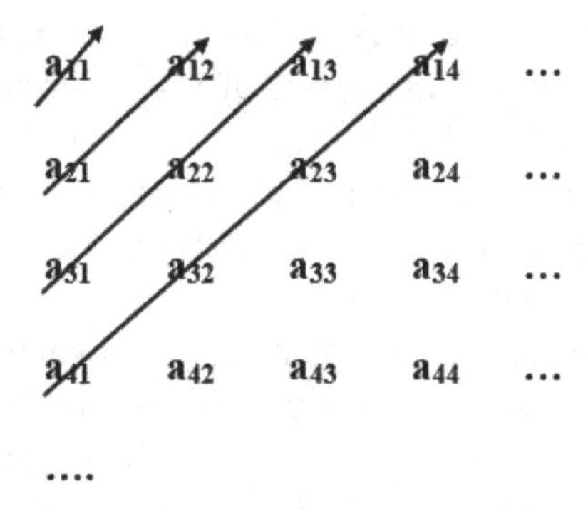

Now construct the sequence as indicated by the arrows, the first 10 terms of which are

$$a_{11}\ a_{21}\ a_{12}\ a_{31}\ a_{22}\ a_{13}\ a_{41}\ a_{32}\ a_{23}\ a_{14}, \ldots$$

Armed with this result, proving the rationals are countable was a slam dunk — the rationals are just the union of the fractions with denominator 1, the fractions with denominator 2, the fractions with denominator 3, etc. — all of which are countable. This proof can be modified to show that the algebraic numbers are countable. So the real numbers consist mostly of transcendental numbers, those numbers which are not the root of a polynomial with integer coefficients, such as e and π.

The rationals, the algebraic numbers, and the transcendental numbers had been considered as possible sets which have a cardinality between aleph-0 and aleph-1. As we have seen above, all of those possibilities flamed out, and mathematicians were left to wonder — is there a set of real numbers which contains the integers, but whose cardinality was neither aleph-0 nor aleph-1?

This question was reframed as the Continuum Hypothesis, known as CH, which states that the answer to the above question is NO — there is no set of real numbers which contains the integers whose cardinality is neither aleph-0 nor aleph-1. Many great mathematicians, including both Cantor and Hilbert, took unsuccessful shots at proving that the Continuum Hypothesis was true (or false). It turned out that they were barking up the wrong tree.

The Axioms of Set Theory

Many mathematicians — myself included — think of set theory as a framework within which we formulate many questions of interest, but we don't really stop to think about what the axioms of set theory really are. When I say "myself included," I can give an example. The first half of my research career focused on mathematical objects known as Banach spaces. I could tell you all the axioms that a Banach space had to satisfy — but I couldn't tell you a single axiom of set theory. I never really needed to know it. I could just say something like "Let E be the set of linear transformations from a Banach space X to a Banach space Y" and move on from there, comfortable in the assurance that there was such a set E and I knew how to play around with it.

But mathematical logicians had dug more deeply into set theory — a LOT more deeply. First of all, they were interested in coming up with a consistent axiom set. A set of axioms is consistent if you can't derive contradictions using it. Nobody in their right mind wants to work with an inconsistent set of axioms — for anything.

By the early 20th century, an industry-standard set of axioms had arrived for set theory, known as ZF for its formulators, Zermelo and Fraenkel. There is an additional axiom, the Axiom of Choice, which is so intriguing I'll discuss it separately a little later; ZF with the Axiom of Choice appended is known as ZFC, and most mathematicians work within its framework. I'm guessing very few actually know the axioms of ZFC; I certainly don't. It's sort of like being able to drive a car without knowing how an internal combustion engine works.

However, two amazing results were to surface a quarter of a century apart. The first was due to Kurt Gödel [4], one of the most brilliant — and troubled — mathematicians of the 20th-century. Gödel was an Austrian, but spent much of his life at the Institute for Advanced Studies in Princeton, NJ. He had the misfortune to have a close friend assassinated, after which he developed the belief that people were trying to poison him. He would only eat food prepared by his wife, and when she was hospitalized for six months, Gödel starved himself to death.

In 1938, Gödel showed that if ZF were consistent (and nobody believes otherwise), then if one adjoins the Axiom of Choice and the Continuum Hypothesis to ZF, it remains consistent. In the language of mathematical logic, if ZF is consistent, so is ZFC + CH.

A quarter of a century later, Paul Cohen of Stanford University, using a totally new proof technique, was able to show that if ~CH denotes the negation of CH (in other words, there is a set of real numbers containing the integers whose cardinality is neither aleph-0 or aleph-1), then ZFC + ~CH is also consistent! In other words, there are two different sets of axioms — ZFC + CH and ZFC + ~CH — and neither leads to a contradiction. Logicians say that CH is independent of ZF.

Even though I had — and have — almost no familiarity with these results, they were to have a significant impact on my life. I had worked for almost a decade, along with other mathematicians, on a major unsolved problem in my area of research. In 1979, I took a five-year break from teaching and research, and when I returned, it had been shown that the problem on which I worked had a different solution depending on whether you decided to adopt or reject CH! The problem which had occupied me for so long turned out to be neither true nor false, but dependent upon which consistent axiom system I chose to work in!

I found this extremely disconcerting. I was willing to accept that there were mathematical propositions that mathematicians would never be able to settle, either because they were too difficult or because they were undecidable. Gödel had shown that one could formulate undecidable propositions — propositions which could not be proven true or false — within any reasonably complex axiom system. I could stomach working on an undecidable proposition — the math research equivalent of drawing dead in poker. But working on something which could be either true or false was not something I was comfortable with, and so I switched into a totally different area of research. You'll see some of that in the final chapter.

The Axiom of Choice

The Axiom of Choice is simple to state — consider the collection of all non-empty subsets of a given set. Here's the Axiom of Choice — it is possible to choose one item from each subset. This seems hardly worth stating — if you have a collection of bags, each of which contains stuff, you go to each bag, reach in, and grab something.

You can certainly do this if there is a rule specifying how you grab stuff. For instance, if you have any number of bags containing positive integers, you can grab the smallest positive integer in each bag — because we know how to find the smallest positive integer in a collection of positive integers.

But Bertrand Russell gave a really seductive way to see the problem with the Axiom of Choice. Suppose you have a collection of pairs of shoes. You can always pick the left shoe from a pair. But if you have a collection of pairs of socks, assuming that the socks are brand new and identical, how do you pick one from each pair? You need the Axiom of Choice to do it.

So consider the following problem — given the collection of non-empty sets of real numbers, how do you specify a rule which tells you how to pick one from each set? Try it — but be warned, this is a problem with no constructive solution.

The Axiom of Choice has some surprising consequences. Perhaps the most famous — or infamous — is the Banach-Tarski Theorem, which says that given any solid sphere, it is possible to break it up into a finite number of pieces and then reassemble them into two spheres, each with the same volume as the original sphere. This mind-boggling result — which no one, mathematicians included, believes until they see the proof, comes about because the pieces aren't really solid, but are collections of points, sort of like swarms of drops in a cloud.

The Axiom of Choice also bears on a problem which interested me ever since I fell in love with PMI — the Principle of Mathematical Induction. PMI works great for countable sets — sets of cardinality aleph-0. But is there a way to do something equivalent for sets of any cardinality?

Transfinite Induction

We can restate the Principle of Mathematical Induction in the following form: if we have a collection of propositions $P(n)$ for each positive integer n, suppose we can prove that if $P(k)$ is true for all $k < n$, then $P(n)$ is true. Then $P(n)$ is true for all integers n.

Set theorists have devised a way to generalize the positive integers that may seem more natural than the idea of cardinal numbers. Although the set $N = \{n | n \text{ is a positive integer}\}$ and the set $E = \{n | n \text{ is an even positive integer}\}$ have the same cardinality — as was shown when a countable collection of guests requested rooms in Hilbert's Hotel — in some sense the set of all positive integers is larger than the set of all even positive integers. After all, the set of all positive integers not only contains the set of all even positive integers, it also contains the odd positive integers.

The way to incorporate this into an extension of the "greater than" property of the integers is called the ordinal numbers. It was developed by — surprise — Georg Cantor, and is one of those concepts that requires a fairly heavy infrastructure to support it. If you're interested in the details, you can find them elsewhere, but the key idea looks something like this — there are a bunch of different ordinals corresponding to different countable sets, and any two of them are either equal or one is greater than the other. And not just countable sets, you can do the same thing for different sets of any cardinality.

What the integers are to the Principle of Mathematical Induction, the ordinals are to Transfinite Induction. Transfinite Induction goes as follows: suppose we have a collection of propositions $P(\alpha)$ for every ordinal α (BTW, this is a LOT of propositions). If we can prove that assuming $P(\beta)$ is true for every ordinal $\beta < \alpha$ enables us to prove that $Pi\alpha)$ is true, then $P(\alpha)$ is true for all ordinals α.

Transfinite Induction is nowhere near as useful as PMI, in part because you just don't find collections of propositions $P(\alpha)$ for every ordinal α. I'm willing to guess that upwards of 90% of the math papers that have been written never even mention the word "ordinal." I know none of mine did. But there is a need for some sort of proof technique when dealing with sets of unknown — and possibly large — cardinality. There are a number of different propositions equivalent to Transfinite Induction, but the workhorse in my area of math — as well as lots of others — is Zorn's Lemma.

Zorn's Lemma

There's some infrastructure needed to support Zorn's Lemma — but not a whole lot. If you want some idea of how math has evolved, this might be a good way to sample it. In fact, from here on out throughout the end of this chapter, all you really need is the ability to comprehend mathematical notation, a willingness to puzzle through a VERY abstract argument (although not a difficult one) — and maybe a stiff drink to get you past the rough spots.

First, some definitions. A *poset* E (shorthand for "partially ordered set") is a set E with a method of comparing two members of the poset. I'll use the usual less-than-or-equal-to symbol, $\leq$, for the method of comparing. In the following three conditions, x, y, and z all belong to E. Informally, we can think of the statement $x \leq y$ as x is smaller than or equal to y — or y is larger than or equal to x.

(1) $x \leq x$ for all x in E

(2) If $x \leq y$ and $y \leq z$, then $x \leq z$

(3) If $x \leq y$ and $y \leq x$, then $x = y$

For numbers, less than or equal to is a partial order. Moreover, for numbers one also has the Law of Trichotomy, any two numbers are either equal or one is less than the other. For sets, subset inclusion is a partial order, but it is certainly not true that if two sets are unequal, one must be a subset of the other. We say x and y are comparable if either $x \leq y$ or $y \leq x$, and we say z is a maximal element for a subset E of the poset if $z \leq u$ implies $u = z$ (in other words, there is no element in E larger than z).

That wasn't so hard, was it? A couple more definitions, and we're ready to go. A chain C in E is a subset of E such that every two elements of C are comparable. We say b is an upper bound for a chain C if for every x in C, $x \leq b$.

Now would be a good time for that stiff drink, because even though the definitions aren't all that difficult, there's some rough sledding ahead.

Zorn's Lemma is simple to state. It says that if E is a non-empty poset in which every chain has an upper bound, then E has a maximal element. Zorn's Lemma does not apply to the integers, because the set comprising all the integers is a chain with no upper bound. However, the set of all subsets of X is a poset to which Zorn's Lemma applies, because the union of all the sets in a chain is an upper bound for the chain.

So let's actually prove something with Zorn's Lemma. We'll go after REALLY big game — the Axiom of Choice. Let X be a set. We'll show that there is a function f defined on X such that for any non-empty subset Y of X, $f(Y)$ belongs to Y.

We're going to define a poset whose elements are pairs (Y,f), where Y is a subset of X and $f(Y)$ belongs to Y. We define $(Y,f) \leq (Y',f')$ if both Y is a subset of Y' and $f'(y) = f(y)$ for all y in Y. This poset is non-empty because for any single point x in X, we just define $f(\{x\}) = x$; clearly $f(\{x\})$ is in $\{x\}$.

Now let C be a chain in this poset consisting of a bunch of pairs (Y,f). Let Y_0 be the union of all the Y in the pairs that make up the chain C. If S is any set on which an f in one of the pairs of the chain C is defined, then all the f in the pairs of the chain C that are defined on S have the same value, so we can define $f_0(S) = f(S)$ without having to worry which f from one of the pairs in the chain C we are using. Clearly (Y_0, f_0) is an upper bound for the chain C. So by Zorn's Lemma the poset has a maximal element, which we'll denote by (V, g).

But $V = X$, because if there were a point x in X but not in V, we could simply define $g(x) = x$, and then $(V \cup \{x\}, g)$ would be larger than the maximal element, and we're done.

This is one of those proofs that, if I were a magician performing a sleight-of-hand illusion rather than a mathematician, I might say, "At no time did the fingers leave the hand." Unlike the PMI proofs, where we actually did computations and stuff, this proof has the feeling of word-and-symbol manipulation rather than an actual proof. But it is a proof, and one that mathematical logicians accept — and they're the people who know best.

I didn't use Zorn's Lemma much in my research, as most of what I did had a much more "constructive" feel. I'll conclude this chapter by recalling the classic joke I mentioned earlier about the mathematician who showed there was enough water to extinguish the fire but didn't bother to do so. It may seem that existence proofs are not especially useful, because they don't tell you how to do something. But they are important, because they tell mathematicians that they are not drawing dead on a particular problem, and that at least there is a solution out there to be found.

And that's especially important when it comes to differential equations. I am in awe of differential equations, to me they are the language of the Universe, because so many processes can be described by them. And if we have a process that we believe is described by a differential equation, it's nice to know that there is a mathematical solution somewhere — because that reinforces our belief that many of the important processes of the Universe function in a way that we can comprehend.

Bibliography

[1] A. Clarke, *The Sentinel* in *Expedition to Earth.* New York, NY: Ballantine Books, 1953.
[2] MacTutor. Online at https://mathshistory.st-andrews.ac.uk/Biographies/Cantor/.
[3] Wikipedia. Online at https://en.wikipedia.org/wiki/11001001.
[4] MacTutor. Online at https://mathshistory.st-andrews.ac.uk/Biographies/Godel/.

Chapter 11

Seduced by Computers

This book is being written on a computer — as is almost everything else these days. The computer I'm using consists of just a keyboard and monitor (of course, with the computer built in), and it's about 10% of the size of the IBM 1620 (picture below) which I programmed to put myself through graduate school. My home computer has millions of times the memory of the 1620, runs thousands of times faster — and is about 10% of the fun as well.

When you sat in front of a 1620, it was like being in command of the starship *Enterprise*. The lights in the blue and gray panels (the panels appear dark gray and light gray in the picture) would blink on and off, indicating the progress of the computer program — what it was doing (the "op" — for operation — code) and where it was doing it. You had to pay attention to the blinking lights, because every so often the computer would enter an infinite loop from which there was no escape. A simple infinite loop looks like

Address	Instruction in that Address
A	Go to B
B	Go to A

You could tell when a computer had entered such a loop because the lights on the panel would blur rather than blink. So you had to stop the program manually (see those switches on the front?) and debug it to find out what caused the error.

This was one of the first problems encountered in the early days of computer programming and led to the Halting Problem. Computer scientists wondered if it was possible to write a diagnostic program which would use as its input a computer program P and input data D — and without doing the computation of the program, simply tell you whether the program would halt or loop.

The individual who solved the Halting Problem — and by an exquisitely insightful proof which can be understood without requiring any mathematical background — was Alan Turing [1]. Alan Turing is possibly the only 20th-century mathematician known outside the generally small circle of mathematicians and scientists — and that's because of three events.

The first of these events was heading the British team of decryption experts headquartered at Bletchley Park during World War II. This team was tasked with breaking the codes constructed by ENIGMA — the German automatic encryption and decryption device used for transmitting military and state secrets. This effort has spawned numerous books, television programs, and motion pictures.

The second event was the Turing Test. Turing proposed that a machine exhibits artificial intelligence if under certain conditions, its answer to specific questions were indistinguishable from the answers that would be given by a human being. Yes, if asked to add 2 and 2, the machine would give an answer indistinguishable from the one given by a human being, but the Turing Test consists of subtler questions. The Turing Test was at one time called the Imitation Game — and there's a movie about this with that title.

The last event for which Turing is known is the saddest. Despite his heroic contribution to the effort to win World War II, Turing was never recognized for this in the list of honors, such as knighthoods, awarded by the British government. Instead, the British government chose to prosecute Turing for homosexual activities. Given the choice of prison or chemical castration, Turing chose the latter — and the consequences of this choice were so catastrophic that he felt it necessary to commit suicide. He thus joined a list of notables from all areas of human endeavor whose achievements were celebrated only after — and to some extent, because — they chose to end their own lives. We can only imagine what these people might have accomplished had their lives not been so sadly cut short.

You'll notice that the solution of the Halting Problem isn't listed in those achievements — but it was certainly one which caused Turing to be thought of as a candidate to head the team at Bletchley Park. Turing's solution to this problem was simple and elegant (the essence, IMHO, of a seductive solution) — and even though the problem was thought to lie in the realm of mathematics, Turing solved it by a purely logical argument about computer programs.

The Halting Problem could be stated as follows: is there a computer program D, which uses as its input another computer program P and the input I to the program P, that can determine whether P, given I as input, halts or loops?

Turing's proof starts by assuming that there is such a program D. D indicates whether P, given input I, halts or loops by doing exactly the same thing — if P, given I, halts, so does D, and if P, given I, loops, so does D. We now construct a new program, which we call D′, which examines what D does and does the exact opposite. If D loops, D′ halts, and if D halts, D′ loops.

Now let's use D′ as the input to D. If D determines that D′ halts, D is supposed to halt. But hold on — if D′ halts, then by construction of D′, D is actually looping! Similarly, if D determines that D' loops, D is supposed to loop. Wait a minute — if D′ loops, then by construction of D′, D actually halts! In other words, when confronted with D, D′ always does the wrong thing. Therefore no program can be constructed which determines whether an arbitrary program and given input will halt or loop.

Computers today are incredibly sophisticated, but everything a modern computer can do, a Turing machine can also do. A Turing machine is an abstract version of a computer consisting of an infinitely long tape with a series of cells on it, and a read-write implement, called a head, which moves back and forth on the tape one cell at a time. The head has a finite number of operations it can perform — usually fairly small. Nonetheless, this apparently primitive machine can be "programmed" to do all the things any computer ever built can do — so in a sense Turing is to computers what Edison was to electric lights.

Chaotic Phenomena

Mathematicians, scientists, and engineers welcomed the arrival of the computers that electronics had made possible. There are numerous problems in all of these disciplines that do not admit exact solutions, but for which there are procedures which arrive at approximate solutions. These procedures previously had to be carried out manually or with the aid of slide rules or mechanical computational devices — and these were limited in both speed and accuracy. The computer enabled orders-of-magnitude improvements in both speed and accuracy.

One such area was weather forecasting. The equations governing the behavior of the weather have been well known for more than a century — but they are partial differential equations, which are notorious for their intractability. In 1960, Edward Lorenz, a meteorologist at the Massachusetts Institute of Technology, had managed to obtain the use of a computer to obtain solutions to these equations.

One day, he needed to take a lunch break, and decided that when he returned, he would start the computer from a convenient intermediate point in the computations that had been reached earlier. He jotted down the values of the key variables, which the computer had calculated to six decimal places. He then made one of the most serendipitous "accidents" in the history of mathematics. Rather than jot down the six-decimal place values of the variables, he rounded them off to three decimal places, figuring that it wouldn't make much difference.

He had, however, kept a record of what the computer did before lunch, when it was working with the six-decimal place values. On returning from lunch, he typed in the three-decimal place rounded values and restarted the computer. A little while later, he noticed that the values he was now obtaining had started to differ from the values he had obtained prior to lunch, and shortly thereafter the solutions he was obtaining after lunch looked nothing like the previous solutions.

Lorenz figured there had been some mistake — either he had typed in values incorrectly, or somehow the computer program had somehow scrambled the computations and weren't solving the original problem. So he checked and rechecked — he had typed in the correct three-decimal place values, and there was nothing wrong with the computer program. The conclusion was inescapable — the apparently trivial differences between three-decimal place values and six-decimal place values had a disproportionate effect on the computations. Lorenz — and the computer — had discovered the existence of chaotic phenomena.

The Traveling Salesman Problem

In a later paper, Lorenz described formally the phenomenon he had discovered, which he termed extreme sensitivity to initial conditions. Informally, it became known as the Butterfly Effect, as Lorenz gave the example of a butterfly flapping its wings in Hawaii would later result in a tornado in Oklahoma. [2]. It's relatively easy to present the idea of sensitivity to initial conditions by using one of the most famous unsolved problems in mathematics, which goes by the name of the Traveling Salesman Problem.

A salesman starts from his home city, and has to visit towns A, B, and C, after which he returns home. The problem asks us to minimize the total distance traveled in doing so. We'll discuss the difficulties inherent in finding a solution fairly shortly, but there's a straightforward algorithm for constructing a travel plan — simply visit the nearest city that has not yet been visited. This algorithm is known as the Nearest Neighbor algorithm, and the following example will show how sensitive it is to small changes in the parameters.

Let's look first at the following mileage table.

	Home	A	B	C
Home	0	100	101	180
A	100	0	120	190
B	101	120	0	130
C	180	190	130	0

The nearest city to Home is A, so we visit it first. B is closer than C to A, so we next visit B. The only unvisited city is C, and from there we go Home. The total distance traveled is $100 + 120 + 130 + 180 = 530$ miles.

As you can see in the previous table, the distance from Home to A is 100, and from Home to B is 101. Let's switch these two distances while leaving all other distances intact. The new distance table is

	Home	A	B	C
Home	0	101	100	180
A	101	0	120	190
B	100	120	0	130
C	180	190	130	0

Now the nearest city to Home is B, so we visit it first. A is closer than C to B, so we go there next. We now proceed to C, and then home. The total distance traveled is 100 + 120 + 190 + 180 = 590 miles. A very small change in a couple of numbers in the table has produced a significant difference in the result produced by the Nearest Neighbor Algorithm. Possibly not on the order of the Butterfly Effect, but at least it gives some idea of the phenomenon.

Easy, Hard, and Nasty Problems

The speed of computers nowadays is staggering. I have an all-in-one I purchased a couple of years ago for less than $1,000; it can go through 1,000,000 add loops in 0.02 seconds — and the language I used to program it is nowhere near optimal in terms of speed. The computer brought the possibility of solving problems which had previously been considered out of reach — and with this new ability a deeper scrutiny of exactly what was retired on the computational front to solve problems.

Roughly speaking, there are three categories of computational difficulty — easy, hard, and nasty. So let's look at an easy problem first. Suppose you need to sort an unordered list of names and put it in alphabetical order. One algorithm for doing this — not the best — is to take each name from the list in its current order, go through the sorted list one at a time from beginning to end (in other words, from A to Z), and see where to place the name. You take the first name from the unsorted list and put it in the sorted list. You then take the second name, compare it with the first, and put it either ahead of the first name or behind it. You then take the third name, compare it with the first sorted name, and then the second sorted name (if necessary) — and so on. The computation depends upon how many comparisons you have to make, and if you already have 17 names in the sorted list, at worst you'll have to make 17 comparisons.

So, if the unsorted list contains N names, the worst-case scenario is that you will have to make 1 comparison for the second name, 2 comparisons for the third name, 3 comparisons for the fourth name, and so on through $N - 1$ comparisons for the Nth name. That means a worst-case total of $1 + 2 + + (N - 1) = N(N - 1)/2$ comparisons. Roughly speaking, the number of computer operations required this sorting algorithm increases as the square of the number N of names on the list. There are better algorithms, but let's stick with this one for the time being.

Let's suppose we have an algorithm such as this, in which the number of computational steps increases as the square of the number of items to be processed. Increasing from 10 to 11 items increases the amount of computation by a factor of $(11/10)^2 = 1.21$, an increase of 21% — but increasing from 100 to 101 items increase the amount of computation by a factor of $(101/100)^2 = 1.0201$, a mere 2.01%. This characterizes the easy problems — the additional percentage increase in time required per item to be processed decreases as the number of items to be processed increases. This behavior is characteristic of any problem for which the number of computational steps increase as a fixed power of the number of items to be processed — the power is 2 for the sorting algorithm just examined.

Temple of Benares/Tower of Hanoi

Nowadays this problem is known as the Tower of Hanoi, but I first encountered this problem in George Gamow's classic *One, Two, Three ... Infinity.* Gamow described the priests in the Temple of Benares as having 64 golden disks, varying in diameter from largest to smallest. The center of each disk was a small hole whose diameter was slightly larger than the diameter of each of three identical spikes, each of which was made of diamond. Initially, the disks were placed one on top of another on one of the three spikes, with the diameters of the spikes increasing from the smallest at the top to the largest at the bottom.

The priests were tasked with the job of transferring the golden disks from one diamond spike to another, subject to the following rules: only one disk could be moved at a time, and no disk of larger diameter could be placed on top of a disk of smaller diameter. So let's suppose that the disks initially are on spike A. We'll let disk 1 be the disk with the smallest diameter, disk 2 the disk with the next-smallest diameter, and so on.

Move 1 Move disk 1 from spike A to spike B.

That was easy!

Move 2 Move disk 2 from spike A to spike C.

Move 3 Move disk 1 from spike B to spike C.

Piece of cake. We now have disks 1 and 2 on spike C in the right order.

Move 4 Move disk 3 from spike A to spike B.

Move 5 Move disk 1 from spike C to spike A.

Move 6 Move disk 2 from spike C to spike B.

Move 7 Move disk 1 from spike A to spike B.

At this stage, you might notice that the number of moves required was 1 for disk 1, 2 for disk 2, and 4 for disk 3. Having seen this pattern before, it shouldn't surprise you that the number of moves is going to double each time you add the next disk. So, to transfer all 64 disks to another spike will require $1 + 2 + \cdots + 2^{63} = 2^{64} - 1$ total moves.

Assuming the priests at the Temple of Benares are sufficiently nimble that they can move one disk per second, it will take $2^{64} - 1$ seconds to do this. There are about 31,500,000 seconds in a year, so the total time required is close to 6 trillion years.

There's a wonderful story by Arthur C. Clarke, *The Nine Billion Names of God* [3], in which a similar set of priests, faced with a roughly similar task (printing out the different names of God), decide to buy a computer to expedite the process. We can write a computer program which mimics what the priests in the Temple of Benares are doing — my computer can do on the order of 50 million transfers per second, but even so it will take it about 117,000 years.

The fact that each successive disk doubles the computational time means that time required goes up exponentially with the number of items being processed. Whether you increase from 10 to 11 disks or from 100 to 101 disks, the computational time required doubles. This is the realm of the hard problems.

The Traveling Salesman Problem Returns

Let's suppose the traveling salesman must visit n cities before he returns home. He has a choice of n cities to visit first, $n - 1$ choices for the next city, $n - 2$ choices for the third city, etc. As a result, there are $n!$ different routes that the traveling salesman could take. Without doing any sophisticated analysis, in order to be sure of finding the shortest route, the computer would have to examine the total distance on each of the $n!$ possible routes.

And now we've entered the realm of the nasty problems. Increasing the number of cities to be visited from n to $n + 1$ requires computing total distances for $(n + 1)!$ routes. Increasing the number of cities from n to $n + 1$ represents a fractional increase of $(n + 1)!/n! = n + 1$ in the computation time. So increasing from 10 to 11 cities requires 11 times the computational time, and increasing from 100 to 101 cities requires 101 times the computational time.

Just because the computational time increases by leaps and bounds doesn't mean there isn't some Alexandrian sword to cut through the Gordian knot. And, interestingly enough, although we'd like to be able to find an algorithm which enables us to find the shortest route in an appropriately short period of time, there's one extremely important problem — to you, me, and everyone on Earth — that we'd prefer to learn that there is no Alexandrian sword available. That problem is factorization — yes, what you did in elementary school and what we touched on when we discussed prime numbers — because the difficulty of factorization lies at the heart of what keeps our computer passwords safe.

Practically every computer password today is encrypted via the RSA (Rivest–Shamir–Adleman) Encryption algorithm [4]. You may have seen *The Man Who Knew Infinity*, a 2015 movie about the great

Indian mathematician Srinivasa Ramanujan. As we saw in Chapter 2, Ramanujan's genius was discovered as the result of a letter he wrote to the English mathematician G. H. Hardy, a man whose life was ostensibly so mundane that he described receiving that letter as the most romantic event of his life. Nearing the end of his life, Hardy wrote a book (*A Mathematician's Apology* [5]) in which he essentially said that his entire life had been spent in the pursuit of what he was convinced was the totally useless beauty of pure number theory — but if artists should be allowed to pursue totally useless beauty, why shouldn't mathematicians? Beauty, after all, is in the eye of the beholder.

But utility isn't. Hardy died some 30 years too soon, for his life's work touches practically everyone today. Hardy spent some of his time studying how difficult it was to factor large numbers that were the product of two prime numbers p and q, and that difficulty underlies the way public key — private key encryption (such as the RSA algorithm) works.

Let's say someone wants to send a private message to Alice. Alice has selected two large prime numbers p and q, and constructs two keys — a public key based on the product pq, and a private key based on the number p. The RSA algorithm enables anyone to use the number pq (the heart of the publicly-available public key) to encrypt a message to Alice (this encryption is based on the public key constructed from pq), but the decryption is based on the private key constructed from p. The exact details are a little wonky, but what keeps our passwords and bank accounts safe involves what Hardy thought of as the totally useless beauty of pure number theory. And if that isn't irony, what is?

So there are some nasty problems for which we'd love to find a quick solution — such as the Traveling Salesman Problem — and some nasty problems that we hope lie forever out of reach, such as the difficulty of factoring large numbers that are the product of two primes. But, surprisingly enough, the two are either equally solvable — or equally intractable. In the early 1970s, Stephen Cook, a mathematician at the University of Toronto, showed that there was a large class of nasty problems with the following property — if you could find a quick solution for one, then you could find a quick solution for all. Once again, the details here are a little wonky, but the key idea is that all of these problems are either (roughly) equally difficult or equally easy.

Half a century later, we still don't know which. But there may be a way to bust these problems wide open using something that never would have occurred to George H. Hardy — or anyone else living in the first half of the 20th century. They didn't have computers back then — but they at least could envision computers. What no one envisioned was a quantum computer — and quantum computers have the potential to supply the Alexandrian sword. There are difficulties to be surmounted, but progress is being made. And because the threat of being able to break passwords using quantum computers is a real one, there is also considerable work being done on how to use those same quantum computers to construct encryption schemes that even quantum computers themselves can't break.

Monte Carlo Simulations

I was a computer programmer back in the Stone Age of computer programming. For the first few years of my computer programming career, an instruction to add two numbers looked something like 22 14108 08116 — and you had to take specialized courses to learn how to write and debug computer programs written like this. But there was a great demand for computer languages which could be easily learned by the scientists and engineers eager to use computers on their own projects, and with the advent of computer languages such as FORTRAN and BASIC, an add instruction looked like $X = Y + Z$ – certainly a lot easier to read and a lot easier to write.

But I don't believe those early computer languages contained a random number generator. In 1955, the Rand Corporation published a book with the catchy title *A Million Random Digits* [6]. I don't think it made any high-profile best-seller lists, but almost everyone who wanted to do Monte Carlo simulations had a copy.

What, you may ask, is a Monte Carlo simulation? It's a mathematical technique, invented by Stanislaw Ulam with the help of John von

Neumann — and the story of its discovery is yet another example of the importance of serendipity. While working on problems associated with the development of the atomic bomb, Ulam came down with a cold. Bedridden, his only entertainment was playing solitaire — the standard version known as Canfield which used to be bundled with Windows and is now an easily-obtainable app for your cell phone. Ulam wondered how often the game of Canfield ended in success — all the cards were piled on the four aces — and tried to compute it using mathematics. This problem turned out to be much too difficult, and Ulam had a brainstorm — why not model the game on computer and use random numbers to deal the cards?

Because this arose in conjunction with Ulam's work on the atomic bomb, it quickly became classified. Just as the code name *Manhattan Project* was given to the effort to develop an atomic bomb, the code name *Monte Carlo* was given to this approach to problem solution — because Ulam's uncle, possibly the black sheep of the family, would borrow money from his relatives to gamble at the casino in Monte Carlo. This method of approaching problems quickly showed its worth in many different areas of mathematics, science, engineering, and finance.

Here's an easy example of how Monte Carlo methods can be used to come up with an estimate for the number π. Imagine that we have a square of side 1 with vertices at (0, 0) and (1, 1) in the xy-plane. Use a uniform distribution to scatter points in this square. Let N be the total number of points in the square, and k the number of points at distance less than 1 from the origin. Points at distance less than 1 from the origin lie inside a quarter-circle of radius 1, the area of this circle is $\pi/4$. So the number k/N will be really close to $\pi/4$ if N is large, and thus provides a good approximation to π.

I don't think early versions of FORTRAN and BASIC had random-number generators as part of the language, but now they do. Over the years I've derived a lot of pleasure — as well as some additional income — devising Monte Carlo simulations to solve problems for which strictly mathematical solutions would be either difficult or impossible.

A true random number generator requires a randomizing device, such as rolling a fair die (I'm not sure how the folks at Rand came up with their million random digits). Nowadays computers use pseudo-random number generators, which are algorithms for generating sequences that, to the naked eye, look like a sequence of random numbers. The current industry-standard pseudo-random-number generator — the Mersenne Twister [7] — is an algorithm developed in 1997 by Makoto Matsumoto and Takuji Nishimura. It is based on the Mersenne prime $2^{19937} - 1$, and passes many of the important statistical tests for randomness.

Making a Name for Yourself — with the Aid of a Computer and This Book

Although Edward Lorenz, the discoverer of chaotic phenomena, did not have the advantage of owning this book (possibly because his work was done half a century before this book was written), he did own a computer. Nowadays, it is possible to purchase a computer that is as far beyond Lorenz's computer as the starship *Enterprise* is beyond the Wright Brothers' monoplane, and probably for a lot less than it cost to build and test the Wright Brothers' monoplane. You can also download free versions of BASIC and learn them reasonably quickly, so you can do everything that I can do — and possibly you already know more sophisticated computer languages than I do, so you're way ahead of me.

Lorenz made his name by using the computer to discover something that the mathematical world had not yet seen, and thus became famous (at least within that limited world). Throughout this book, there are problems which can be attacked by amateurs armed only with a computer; I'll discuss a couple of the early ones and let you scout out the others.

In Chapter 2, we discussed the Kaprekar Numbers. The Kaprekar sequence starts by taking any four digit number, writing the digits in ascending and descending order, subtracting the smaller from the larger, and continuing the process. Eventually you hit 6,174, and the process loops from there. I don't know for sure, but I'd guess Kaprekar found this by playing around with an assortment of rules using a computer. Probably some of those rules didn't do anything interesting — but then he found one that did.

And, as I said in Chapter 2, nothing is stopping you from doing the same thing. Here's the rule I talked about in Chapter 2. I don't know how it works, but I'm just using it as an example. Take any four digit number, and move the last digit to the front of the number. For instance, take the number 2021 — the year in which I'm writing this book. If you move the last digit to the front of the number, you get 1202. Subtract the smaller number from the larger number, getting 919 (which we write as a four-digit number 0919). Move the last digit to the front of the line, and continue the process. This rule is so easy that you can do it in Excel rather than a formal computer language like BASIC.

Come up with something interesting, and who knows what will happen? Maybe you'll get a collection of numbers named after you, like Kaprekar.

In Chapter 4, we came upon a problem that might conceivably be solved by a 6-year-old. You have a square 201 inches on a side, and six 100×100 inch squares. Can you cover every point on the 201 inch square by using the 100 inch squares, but without cutting those squares? As of this date, nobody knows — but if it can be done, you might be the one to do it, and you can do so by writing a computer program that uses a Monte Carlo simulation. Here's how to go about it — I'll leave the details to you.

Think of the 201 inch square as being positioned in the first quadrant of the xy-plane, with one of the vertices at the origin. Any one of the 100 inch squares is positioned by knowing the location of two vertices that lie on a diagonal — these two vertices will be a distance $100\sqrt{2}$ in the xy-plane. Use the random number generator to position the six 100 inch squares by knowing the co-ordinates of two vertices lying on a diagonal as described in the previous sentences. Now see if every other point is covered.

Of course, finding this solution requires first of all that there is a solution (and currently, nobody knows, which is why you're doing this). Second, you want to be sure that your computer HALTS and prints "Eureka!" — or some similar exclamation of triumph — if and when it finds it. So the debugging process for this particular computer program involves trying to do this in situations where you know there is a solution, and seeing that your computer program does indeed HALT and print "Eureka!" when it finds a solution.

This book contains other problems that a computer might attack — but don't let this book limit you. There's a universe of problems waiting to be explored, and the computer is a tool of exploration like no other ever invented. Happy hunting!

Bibliography

[1] MacTutor. Online at https://mathshistory.st-andrews.ac.uk/Biographies/Turing/.

[2] E. Lorenz, *Deterministic Nonperiodic Flow*. Journal of the Atmospheric Sciences, 1963.

[3] Online at https://urbigenous.net/library/nine_billion_names_of_god.html.

[4] Wikipedia. Online at https://en.wikipedia.org/wiki/RSA_(cryptosystem).

[5] G. Hardy, *A Mathematician's Apology.* Cambridge UK: Cambridge University Press, 1940.

[6] *A Million Random Digits With 100,000 Normal Deviates.* Santa Monica, CA: Rand Corporation, 1955.

[7] Wikipedia. Online at https://en.wikipedia.org/wiki/Mersenne_Twister.

Chapter 12

Seduced by a Few of My Favorite Things

If you're writing a thriller, it should have a slam-bang finish. Unfortunately, that's difficult to do with a book of this type, because it's a collection rather than a single theme. I've struggled trying to decide what to put in this chapter, and finally decided to just go with things I found especially enjoyable over the course of the years. Some of the material is a little advanced, but I'm going to conclude with a theorem that was published in 1955 in *the Pacific Journal of Mathematics* whose proof can be understood by a high-school student who is willing to spend maybe an hour at most to assimilate the preliminaries.

My Favorite Mathematical Joke

If you are contemplating a career as a stand-up comedian, don't even consider telling this joke — unless your audience consists exclusively of people with some familiarity with calculus. Even then, they're not going to laugh hysterically. But, here goes.

Two math teachers are dining at a restaurant, and, as math teachers are wont to do, deploring the generally low level of mathematical knowledge of the world in general and their students in particular. One of them says, "They don't even know what the integral of one over *x dx* is any more." He excuses himself to go to the restroom, and the remaining teacher calls the waitress over.

"I'd like you to do me a small favor. When you come to take our orders, I'm going to ask you a question. I'd appreciate it if you would answer 'the natural log of *x*.' Can you do that?"

"Sure," says the waitress, "The natural log of *x*. No problem." She leaves, and after a while the other teacher returns.

Shortly thereafter, the waitress returns to take their order. The teacher who remained at the table says, "Miss, do you mind if I ask you a question?"

"As long as it isn't too personal, sure," she replies.

"What is the integral of one over *x dx*?" asks the teacher who remained at the table.

"The natural log of *x* plus C," replies the waitress.

Granted, it's subtle, and not a knee-slapper (most mathematical jokes are so dry they're desiccated). I always tell this in first-semester calculus classes as a way to help students remember the constant of integration in indefinite integrals.

My Next Favorite Mathematical Joke

This one is definitely shorter and probably funnier (IMHO), but without the pedagogical ramifications that moved the previous joke into the top spot. It may be apocryphal, but it was related to me as actually happening.

A woman with a young child was studying mathematics. The child asked, "Mommy, what are you doing?"

"I'm studying abelian groups," was the reply.

"Mommy, are abelian groups good to eat?" asked the child.

And thus was conceived **The Fundamental Theorem of Edible Groups** — All edible groups are non-abelian.

One Good Whack

One of the most seductive things about — well, the Universe — is how the explanations of phenomena can be couched in the language of mathematics. I'm not one of those nerdy people who think that math will eventually explain everything — partly because I don't believe everything is explicable, and partly because I don't believe math is capable of explaining everything that is explicable. But I could be wrong.

I've always loved differential equations — it's a source of continual astonishment that so many phenomena, in such disparate worlds as the natural sciences, the social sciences, and finance, are explained by differential equations. In plain English, a differential equation relates how quantities are changing, and the solution to such an equation tells us how the parameters in the differential equation are related to one another. This is what prompted the brilliant French mathematician Pierre-Simon de Laplace, to proclaim that if he knew the position and velocity of everything at this moment, he would know where everything would be at all times in the future [1].

And differential equations supplied me with an answer to a question that had troubled me since childhood. My childhood, BTW, came at an early stage of electromechanical devices, and it was common practice if your radio stopped working, to just turn it off, give it a whack, and then turn it back on. More often than not, this worked. BTW, it still does, I do the same to my computer when it stops working (which thankfully isn't often) — although I'm gentler with my computer than I was with my radio, partly because the computer is a lot more expensive.

Although it is tempting to believe that the whack awakened the slumbering spirit responsible for the proper functioning of the machine, there is a less metaphysical explanation for the efficacy of this procedure in terms of basic differential equations.

As an example, suppose that we have a balky device whose motion $y = y(t)$ is described by the differential equation $y'' + y = 0$. The general solution to this differential equation is $y = A \sin t + B \cos t = C \sin (t + \varphi)$. This is standard sinusoidal behavior, such as might be exhibited by simple alternating current. We can describe its initial position as $y(0) = 0$ and because it shows no interest in doing anything, its initial velocity is given by $y'(0) = 0$. However, when we give it a good whack of magnitude M at time $t = t_0$, the differential equation now becomes $y'' + y = M\delta(t - t_0)$, where $\delta(t - t_0)$ is the Dirac delta function jolted at time $t = t_0$. The solution to this equation is $y(t) = 0$ for $t \le t_0$, $y(t) = M \sin (t - t_0)$ for $t > t_0$. In plain English, nothing happened until we whacked it, but after the whack it started functioning just like it would have normally.

To the Vector Belong the Spoils

Sorry about that, but I've wanted to say this for almost half a century.

Vectors first put in a serious appearance in multivariable calculus, and they do the same thing for three-dimensional geometry that analytic geometry does for two-dimensional geometry. Vectors supply a way to give answers to geometric problems via algebraic computations.

Quantities with magnitude, such as distance and mass, are called scalars. Vectors are quantities with both magnitude and direction. The three classic examples of vectors are directed distance (3 miles headed due East), velocity (40 miles per hour due North), and force (lifting 10 pounds straight up). But what makes vectors so amazingly useful are the two vector products. The reason these are so useful is that each of the two products has an algebraic method of computation, and a geometrical interpretation.

I'll cut to the chase here, but any multivariable calculus text will have the justifications of both the algebraic method of computation and the geometrical interpretation. Although these will normally occur in the third semester of a calculus course, there's no calculus at all used anywhere, so in theory it should be possible for a high-school student who has had analytic geometry to understand it.

We'll assume that $\boldsymbol{v} = v_1\boldsymbol{i} + v_2\boldsymbol{j} + v_3\boldsymbol{k}$ and $\boldsymbol{w} = w_1\boldsymbol{i} + w_2\boldsymbol{j} + w_3\boldsymbol{k}$, where $\boldsymbol{i}$, $\boldsymbol{j}$, and $\boldsymbol{k}$ are the unit vectors in the positive x, y, and z direction respectively. The length of $\boldsymbol{v}$ is $|\boldsymbol{v}| = \sqrt{v_1^2 + v_2^2 + v_3^2}$.

Dot product $\boldsymbol{v} \bullet \boldsymbol{w}$. The algebraic definition is $\boldsymbol{v} \bullet \boldsymbol{w} = v_1w_1 + v_2w_2 + v_3w_e$. This makes the computation of $\boldsymbol{v} \bullet \boldsymbol{w}$ a slam dunk. But it's the geometric interpretation of the dot product that knocks it out of the ballpark (mixing sports metaphors here), as $\boldsymbol{v} \bullet \boldsymbol{w} = |\boldsymbol{v}|\ |\boldsymbol{w}| \cos \vartheta$, where ϑ is the angle between $\boldsymbol{v}$ and $\boldsymbol{w}$. One of the reasons that this is so useful is that it gives an immediate condition for perpendicularity — two non-zero vectors are perpendicular if and only if $\boldsymbol{v} \bullet \boldsymbol{w} = 0$. The reason this is so valuable is because perpendicularity is extremely important, and often difficult to establish through purely geometric arguments. But calculating $\boldsymbol{v} \bullet \boldsymbol{w}$ is really easy.

Cross product $\boldsymbol{v} \times \boldsymbol{w}$. There's a truly ghastly formula for writing out $\boldsymbol{v} \times \boldsymbol{w}$ as a linear combination of the three unit vectors. If you know this, you have no life — or at least, that's what I tell my students. As a case in point, I don't know this. The algebraic computation is best handled not by formula, but by the determinant

$$\boldsymbol{v} \times \boldsymbol{w} = \begin{vmatrix} \boldsymbol{I} & \boldsymbol{j} & \boldsymbol{k} \\ v_1 & v_2 & v_3 \\ w_1 & w_2 & w_3 \end{vmatrix}$$

The geometric interpretation of $\boldsymbol{v} \times \boldsymbol{w}$ is that it is a vector perpendicular to both $\boldsymbol{v}$ and $\boldsymbol{w}$ in such a way that $\boldsymbol{v}$, $\boldsymbol{w}$, and $\boldsymbol{v} \times \boldsymbol{w}$ form a right-handed triple, and that the length of $\boldsymbol{v} \times \boldsymbol{w}$ is $|\boldsymbol{v}|\ |\boldsymbol{w}| \sin \vartheta$, where ϑ is the angle between $\boldsymbol{v}$ and $\boldsymbol{w}$. Here we see a criterion for two non-zero vectors to be parallel; they are parallel if and only if $\boldsymbol{v} \times \boldsymbol{w} = 0$ — and if you thought proving things are perpendicular was hard, proving things are parallel is even trickier — at least, if the only tool you have available is geometry.

Another interpretation of the length of $\boldsymbol{v} \times \boldsymbol{w}$ is that it is the area of the parallelogram which has $\boldsymbol{v}$ and $\boldsymbol{w}$ as adjacent sides. To see how useful this is, consider the problem of finding the area of a triangle in three-dimensional space if all you know is the co-ordinates of all three vertices. I'm not exactly sure how to go about this using run-of-the-mill analytic geometry. Well, I could do it, but every technique I can think of — with one exception — is messy and lengthy. The one exception is to use Heron's Formula, which gives the area of a triangle in terms of the lengths of its three sides. You can use the distance formula between points in three-dimensional space to compute the lengths of the three sides, and then apply Heron's formula, but this requires you to do a lot of really tedious work.

However, if you just pick one of the three points, compute the vectors from that point to each of the other two points, and then take half the length of the cross-product of those two vectors, you're done! It's half the length because the area of the triangle is half the area of the parallelogram. How cool is that?

Dot and cross products handle a host of other problems as well. I'm not going to go into them here, but the basic geometry problems involving lines and planes, and questions of when these objects are perpendicular or parallel to each other, are all handled using dot and cross products. And then, when the time comes to compute quantities involving surfaces, such as the equations of tangent planes or the computation of surface areas, dot and cross products really shine.

And finally, if you look at one of the absolutely stellar achievements in physics — Maxwell's Equations — they combine the two topics I've just discussed — vectors and differential equations. These expressions are so beautiful — simply as an array of graphical symbols — that I feel it adds to the overall seductiveness of the book if I simply write them down.

And so I shall.

$$\nabla \bullet \mathbf{E} = 4\pi\rho$$

$$\nabla \bullet \boldsymbol{B} = 0$$

$$\nabla \times \mathbf{E} = -\frac{1}{c}\frac{\partial \boldsymbol{B}}{\partial t}$$

$$\nabla \times \boldsymbol{B} = \frac{1}{c}\left(4\pi\mathbf{J} + \frac{\partial \mathbf{E}}{\partial t}\right)$$

Equivalence Relations

You don't really hear about equivalence relations until you take courses after completing the calculus sequence, but you've been using them ever since you were old enough to speak.

So what's an equivalence relation?

Suppose you have a set X. Strictly speaking, a relation is a subset E of $X \times X$, the set of all pairs (x, y) where x and y both belong to X, but since it's a subset, we'll just say xRy — meaning x is related to y — if (x, y) belongs to E. There are all sorts of relations — some are simple (such as X being the set of all physical objects, and xRy meaning x weighs less than y), and some are more complicated (such as X being the set of all people, and xRy meaning that x and y have a friend in common).

A relation is an equivalence relation if it satisfies three properties.

(1) Reflexivity — for any x in X, xRx

(2) Symmetry — for any pair x and y both in X, if xRy, then yRx

(3) Transitivity — for any three elements x, y, z in X, if xRy and yRz, then xRz

The relation "weighing less than" is transitive, but neither reflexive nor symmetric. The relation "have a friend in common" is reflexive and symmetric, but not transitive. John and Ellen may have a friend in common, and Ellen and Mary may have a friend in common, but John and Mary may have no common acquaintances, to say nothing of common friends.

Anyway, there are lots of equivalence relations. For instance, looking at the set of cars, if xRy means both x and y have the same manufacturer, xRy is an equivalence relation. Looking at the set of all people, if xRy means both x and y have the same biological parents, xRy is an equivalence relation.

Suppose that R is an equivalence relation on a set x, and that $R[x] = \{y \text{ in } X | xRy\}$. In words, $R[x]$ is the set of all things equivalent to x. Since xRx, x is in $R[x]$ for any x.

The basic fact about equivalence relations is this — for any two elements x, y in X, either $R[x] = R[y]$ or $R[x\} \cap R[y] = \varphi$. If $R[x\} \cap R[y] \neq \varphi$, suppose z belongs to $R[x\} \cap R[y]$. So both xRz and yRz. Then if u belongs to $R[x]$, then xRu. By symmetry, uRx. Since xRz, by transitivity uRz. Again by symmetry, zRy, so using transitivity one more time, uRy. Last call for symmetry, yRu — so by definition, u belongs to $R[y\}$. Therefore $R\{x]$ is a subset of $R[y]$. Starting with "Then if u," switch the letters x and y to conclude that $R[y]$ is a subset of $R[x]$, and so $R\{x] = R[y]$.

Equivalence classes therefore break X up into a collection of disjoint sets, which are called equivalence classes — $R[x]$ is the equivalence class to which x, and anything equivalent to x, belongs. For the "same manufacturer" relation, the equivalence classes are all Chevrolets, or all Hondas, etc. For the "same parents" relation, the equivalence classes are collections of siblings.

You probably won't be surprised to know that equivalence classes of Hondas or siblings are generally not of interest to mathematicians. I'm not sure whether there is a general theorem to the following effect, but here goes: sets with certain mathematical structures admit equivalence relations such that the collection of all equivalence classes can be given the same structure. The list of examples of this is extensive. If N is the set of all integers with the addition operation, define integers x and y to be equivalent if they differ by a multiple of 12. The equivalence classes are integers which have the same remainder when divided by 12, and those equivalence classes can be looked at as the integers modulo 12 with the addition operation. If we think of the equivalence class of 0 being what we think of as the equivalence class of 12, we're looking at the times 1o'clock, 2o'clock,..., 12o'clock. To add 5 hours to 9o'clock, we simply add $5 + 9 = 14$ and subtract 12, getting 2o'clock.

For some unknown (to me) reason, these collections of equivalence classes are defined by using the word "quotient" — quotient groups, quotient rings, quotient spaces, etc. You've already seen their utility by realizing we use these things to tell time.

Mathematical Results with Amusing Names

I'm a sucker for mnemonic devices — some association that helps you remember things. Who can forget ROY G BIV? Well, if you have, or you've never encountered Mr. Biv, it's the colors of the rainbow from outside to inside — Red, Orange, Yellow, Green, Blue, Indigo, Violet. And even though I grew up before FOIL and SOH-CAH-TOA appeared on the mathematical scene, I can appreciate someone trying to make the product of two binomials memorable (First, Outer, Inner, Last) or the trig functions in a right triangle (Sine Opposite Hypotenuse — Cosine Adjacent Hypotenuse — Tangent Opposite Adjacent). I'm pretty sure it's major political incorrectness to remember the Pythagorean Theorem by the overweight Native American women, but Whiskey Tango Foxtrot, the squaw on the hippopotamus is equal to the sum of the squaws on the two adjacent hides.

So here are some mathematical results with associated mnemonics.

The Pigeonhole Principle — I'm guessing this one is not long for this world, as nobody sees pigeonholes any more. But the principle is simple enough, if the number of pigeons exceeds the number of pigeonholes available to house them, some pigeonhole is going to have to house more than one pigeon. Sounds almost trivial, but it's surprisingly useful.

The Chinese Restaurant Principle — I almost stumbled upon this before my 10th birthday. Back in the day, your basic Chinese restaurant would offer a prix fixe meal where you could choose an appetizer from column A and an entrée from Column B. I would always go for the egg flour soup and the bacon wrapped shrimp (I was able to consume unbelievable quantities of cholesterol when younger, those days are sadly gone forever). My father, however, seemed to delight in ordering a different meal every time, and I wondered how long it would take him to try every possible meal. The answer is the product of the number of appetizers times the number of entrees. And that's the Chinese Restaurant Principle — the number of combined choices is equal to the product of the number of separate choices.

I haven't seen a Chinese restaurant offering this in decades. It could be that this went out with 78 rpm records (popular when my Dad and I frequented Chinese restaurants) — or maybe the fact that Los Angeles has more sophisticated Chinese restaurants.

The Hairy Billiard Ball Theorem — You can't comb a hairy billiard ball without getting cowlicks. How can you not love this one? You envision trying to comb a hairy billiard ball by combing it parallel to the equator — and then when you get near the North and South poles, you just can't do it. The mathematical theorem is that there is no continuous vector field on the surface of a sphere.

Why is there no hairy doughnut theorem? Well, because combing the hair parallel to the equator on a doughnut works just fine, showing that there are continuous vector fields on the torus.

The Ham Sandwich Theorem — Toss two slices of bread and a slice of ham anywhere in the Universe. Then get yourself a REALLY large knife, and if you choose the right way to angle the cut, you can put half of each slice of bread and half of the ham on one side of the cut, and the remaining bread and ham on the other.

This theorem comes from a branch of mathematics called measure theory, which we've encountered before but avoided discussing. Remember the Axiom of Choice and the Banach-Tarski Theorem, where we separated a sphere into five clouds of points, moved them somewhere else without changing their relative positions, and reassembled them into a sphere of twice the volume? One can only do this with clouds of points that are not what are called measurable. Here, the Ham Sandwich theorem requires that the three objects (in our case, two slices of bread and one slice of ham) be measurable.

The Ugly Duckling Theorem — Remember Hans Christian Andersen's story about the misplaced egg that hatched into a baby swan amid a flock of ducklings, and the ducklings thought the beautiful swan was an ugly duckling? The idea here is that a swan and a duckling have just as many attributes in common as two different ducklings — one just has to choose the attributes correctly. Unless there is some sort of inherent bias in attribute selection (such as "beautiful" is more preferable to "ugly" than "bigger than a breadbox" is to "smaller than a breadbox"), any two distinct individuals belong to the same number of classes with the same overall value.

Fixed-Point Theorems

Suppose that X is a set, and T is a function from X into X. Such functions are sometimes called self-maps. We say that x is a fixed point of T if $Tx = x$. The best example I know from the real world is the eye of a hurricane — while everything around the eye is transported to a different position by the force of the hurricane, the eye remains still. Everyone who has ever been caught in the eye of a hurricane remarks upon how eerie the experience seems — while all around the eye all hell is breaking loose, it is calm and serene within the eye.

There are three fixed-point theorems I find fascinating. The first is the Brouwer Fixed-Point Theorem, which states that any continuous map of the solid sphere into itself has a fixed point. The theorem is actually far more general than that — it states that any self-map of a closed and bounded convex set in Euclidean n-space has a fixed point. A set is convex if any line segment drawn between two points in the set remains within the set. The unit disk, consisting of all points at distance less than or equal to 1 in the plane, is convex, but its boundary, the unit circle, is not. It's easy to find a self-map of the unit circle which has no fixed point — just rotate the circle 90 degrees clockwise, and every point is at a different location from where it started.

The second is the Banach Contraction Principle. It's easiest to discuss it for self-maps of the real line, a contraction is one for which there is a constant k with $0 < k < 1$ and $|Tx-Ty| \leq k\,|x-y|$. The Banach Contraction Principle states that there is one and only one real number x such that $Tx = x$. Moreover, there is a straightforward procedure for finding x — simply start with any real number r, compute Tr, $T^2(r) = T(Tr)$, $T^3(r) = T(T^2(r))$,.... The sequence $\{T^n(r) | n = 1, 2,...\}$ converges to the unique fixed point of T.

Here's an example of how this works. The function $f(x) = 3 + \sin(x/5)$ is a self-map satisfying the contraction condition; I chose $x = 9.01$ as a starting point; 9.01 is my wife's birthday, and woe betide me if I forget that. I made a simple Excel table, here's the result for the first ten iterates.

x	*f*(*x*)
9.01	3.973391
3.973391	3.713638
3.713638	3.6763
3.6763	3.67078
3.67078	3.669961
3.669961	3.669839
3.669839	3.669821
3.669821	3.669818
3.669818	3.669818
3.669818	3.669818

One of the interesting things about the Banach Contraction Principle is that it can be used to prove that certain differential equations have solutions — and I've stated how I feel that knowing differential equations have solutions is important because it means that, since differential equations describe so many important physical phenomena, we're not drawing dead in trying to understand the Universe

But the theorem with which I intend to bring down the curtain — the one that first appeared in the *Pacific Journal of Mathematics* in 1955 — is the Tarski Fixed Point Theorem [2]. Yes, the same Tarski that showed up in the Banach–Tarski Theorem.

Tarski, like David Blackwell, was another prestigious mathematician at the University of California at Berkeley while I was a graduate student. Like Blackwell, I never heard of him and never crossed paths — if you're a grad student, you generally take courses in algebra and analysis, and then start specializing. Then as now, Berkeley had a number of internationally-celebrated mathematicians in areas different from the ones I pursued. Just as a football fan may have no idea who the champion speed-skaters are, someone studying analysis may have no idea who the great logicians and statisticians are. I could have been drinking a cup of coffee right next to them and never have known.

All the fixed-point theorems of which I am aware occur in specific mathematical objects. The Brouwer Fixed-Point Theorem occurs in compact convex subsets of n-dimensional Euclidean space, the Banach Contraction Principle in complete metric spaces. Tarski's Fixed Point Theorem takes place in complete lattices — so we'd better learn what complete lattices are.

Remember posets from Chapter 10 (not so long ago)? We start with a poset X with partial order $\leq$, but we need a few more definitions.

Let E be a subset of X. An upper bound u for E is an element u in P such that $x \leq u$ for all x in E. Similarly, a lower bound l for E is an element l in P such that $l \leq x$ for all x in E. These are pretty straightforward definitions to understand — upper bounds are as big or bigger than anything in E, and lower bounds are as small or smaller than anything in E. Sets may or may not have upper bounds or lower bounds, depending on a variety of factors. For instance, in the poset R of the real numbers with the usual ordering, the set of all integers (positive and negative) has neither an upper nor a lower bound.

Here's the tricky definition. If a set E does have an upper bound, the least upper bound for E is an upper bound which is as small or smaller than any upper bound. Similarly, if a set E does have a lower bound, the greatest lower bound for E is a lower bound which is as large or larger than any upper bound. We denote the least upper bound of E, if it exists, as LUB E, and the greatest lower bound, if it exists, as GLB E.

There is a stumbling block here that students taking a course in mathematical analysis encounter immediately. Consider the set E of all rational numbers whose square is less than 2. As a subset of the poset R of real numbers, it has a least upper bound — the square root of 2. But as a subset of the poset Q of rational numbers, it does NOT have a least upper bound! The proof, which can be found VERY early in Rudin's *Principles of Mathematical Analysis* — and which provides a preview of coming attractions (or terrors, depending on your POV) — consists of showing three things.

If LUB E exists, it can't be the square root of 2, because the square root of w is irrational, and so isn't in Q.

If $(\text{LUB } E)^2 < 2$, it is possible to find another rational $q > \text{LUB } E$ which is also in E, contradicting the fact that LUB E was an upper bound for E.

If $(\text{LUB } E)^2 > 2$, it is possible to find another rational $q < \text{LUB } E$ which is also an upper bound for E, contradicting the fact that LUB E was the LEAST upper bound for E.

So the existence of LUBs and GLBs is not a straightforward proposition.

We're almost ready for the grand finale.

A lattice is a poset such that any two elements have both a LUB and a GLB. There are two classic examples of lattices. The set of all subsets of a given set S forms a lattice (with $\leq$ being set inclusion), with the LUB of two subsets being the union and the GLB being the intersection. The real numbers are a lattice with the LUB of any two numbers being their maximum, and the GLB of any two numbers being their minimum.

Penultimate definition — a complete lattice L is one for which every non-empty subset of L has both a LUB and a GLB. Since L is a subset of L, its GLB is less than or equal to any element in L, we write $0 = \text{GLB } L$. Similarly, LUB L is greater than or equal to any element in L, and we write $1 = \text{LUB } L$.

The lattice of all subsets of a given set S is complete. Given any collection of subsets of S, its LUB is the union of all those subsets, and its GLB is the intersection of all those subsets.

The lattice of real numbers R is not complete — the set of all integers has neither LUB nor GLB. However, if we just look at the closed interval [0,1], it is complete — LUB and GLB in this lattice are just the usual (for those of you with some background in math analysis) least upper bound and greatest lower bound.

And now for the final definition. Suppose that L is a complete lattice, and T is a self-map of L. We say that T is order-preserving if $x \leq y$ implies $Tx \leq Ty$. These maps are sometimes called "isotone" in the literature, but although it sounds more seductive to say "isotone" than to say "order-preserving," the latter at least gives you an idea of what it means.

Time for the final act — Tarski's Fixed-Point Theorem for Complete Lattices. Every order-preserving map of a complete lattice has a fixed point.

The proof is SO simple it's almost scary, and you'll wonder why it took until 1955 for someone (Tarski) to notice this. Let E be the set of all elements x in L such that $Tx \leq x$. E is non-empty, since $T1 \leq 1$ (every element in L is ≤ 1). Let $z = \text{GLB } E$.

If x belongs to E, $z \leq x$. Since T is order-preserving, $Tz \leq Tx$. But, since x belongs to E, $Tx \leq x$, and by transitivity, $Tz \leq Tx \leq x$. Therefore Tz is a lower bound for E, and so $Tz \leq z$, since z is the greatest lower bound for E.

Since $Tz \leq z$ and T is order-preserving, $T(Tz) \leq Tz$. But that means that Tz satisfies the criterion for belonging to E, and since $z = \text{GLB } E$, $z \leq Tz$. Since both $Tz \leq z$ and $z \leq Tz$, we see that $Tz = z$, and z is a fixed point of T.

I'm guessing that when this theorem was published, there was a collective groan uttered by the world's lattice theorists, accompanied by saying, "Why didn't I think of that?"

And that's one of the reasons that mathematicians find mathematical research seductive — the hope of finding something so simple and beautiful as a theorem like this.

I did a lot of mathematical research — and none of it was anywhere near as elegant or beautiful as this. But there was always a moment when I'd proved something that I thought to myself, "I may be the only entity in the Universe that knows this." And when a result gets published, there's always the hope that someone else may find it beautiful or useful — or maybe both.

And I think that's what Hardy was talking about in *A Mathematician's Apology*. I'd be willing to guess that very few mathematicians feel that what they did with their careers will have anywhere near the impact of what Hardy did — which led to the RSA algorithm which preserves the secrecy of our passwords. But I'd also be willing to guess that almost all mathematicians feel as I do, and as Hardy did — that a life spent teaching and pursuing the beauty inherent in mathematics is a life well spent.

Time to bring down the curtain.

Bibliography

[1] P. de Laplace, *Theorie Analytique de Probabilites: Introduction,* v. VII, *Oeuvres* (1812–1820).

[2] A. Tarski, A lattice-theoretical fixpoint theorem and its applications. *Pacific Journal of Mathematics*, 1955.

Index